Techniques of Weed and Pest Control

Techniques of Weed and Pest Control

Edited by
Jax Bailey

☐ Larsen & Keller
www.larsen-keller.com

Techniques of Weed and Pest Control
Edited by Jax Bailey
ISBN: 978-1-63549-293-4 (Hardback)

▤ Larsen & Keller

Published by Larsen and Keller Education,
5 Penn Plaza,
19th Floor,
New York, NY 10001, USA

Cataloging-in-Publication Data

Techniques of weed and pest control / edited by Jax Bailey.
    p. cm.
Includes bibliographical references and index.
ISBN 978-1-63549-293-4

1. Weeds--Control. 2. Pests--Control. 3. Pesticides. 4. Insecticides. I. Bailey, Jax.
SB950 .T43 2017
632.9--dc23

The publisher's policy is to use permanent paper from mills that operate a sustainable forestry policy. Furthermore, the publisher ensures that the text paper and cover boards used have met acceptable environmental accreditation standards.

Printed and bound in the United States of America.

For more information regarding Larsen and Keller Education and its products, please visit the publisher's website www.larsen-keller.com

Table of Contents

Permissions

Index

Preface

This book explores all the important aspects of weed and pest control in the present day scenario. Pest control is a practice used to control, manage and regulate the harmful pests. Weed control is a part of pest control and is used to control the growth of unwanted injurious and noxious plants called weeds. This text is a valuable compilation of topics, ranging from the basic to the most complex theories and principles in the field of weed and pest control. It aims to serve as a resource guide for students and experts alike and contribute to the growth of the discipline.

A foreword of all Chapters of the book is provided below:

Chapter 1 - Weed control is the attempt to stop noxious or injurious weeds from competing with domestic plants while pest control is the regulation or management of a species defined as a pest, and can be harmful for a person's health. It is an integral part of agricultural sciences. This chapter will provide an integrated understanding of weed and pest control.; **Chapter 2 -** Noxious weeds or injurious weeds are injurious to agricultural crops, natural habitat or humans and livestock whereas an invasive species is a plant or fungus that is not native to a specific location. The following content elucidates about harmful weeds and pests, and gives a brief explanation on beneficial weeds, noxious weeds, invasive species, aphid and western corn rootworm.; **Chapter 3 -** Mechanical weed control technique manages to remove weed physically by injuring or making the growing conditions unfavorable. This chapter discusses the methods of weed control in a critical manner providing key analysis to the subject matter. It elucidates methods such as mechanical weed control, bush regeneration, soil steam sterilization and irrigation. This chapter is a compilation of the various branches of weed control that form an integral part of the broader subject matter.; **Chapter 4 -** This segment carefully elaborates the basic concepts of pest control to provide a complete understanding. Processes and techniques, like mechanical pest control, physical pest control, fumigation, crop rotation and integrated pest management are explained in a critical and systematic manner. This chapter is an overview of the subject matter incorporating all the major aspects of weed and pest control.; **Chapter 5 -** The substances used for attracting, seducing and then destroying any pest is known as pesticides. Pesticides are commonly used for plant protection, while an insecticide is a substance used to kill insects. The aim of this chapter is to provide the readers an in-depth understanding on pesticides and insecticides and their importance in agriculture.; **Chapter 6 -** Plant disease resistance is the reduction of pathogen growth on or in the plant, at the same time plant disease defense is a range of adaptions evolved by plants, which improve their survival. The aspects elucidated in this section are of vital importance, and provide a better understanding on the subject matter.

I would like to thank the entire editorial team who made sincere efforts for this book and my family who supported me in my efforts of working on this book. I take this opportunity to thank all those who have been a guiding force throughout my life.

Editor

Introduction to Weed and Pest Control

Weed control is the attempt to stop noxious or injurious weeds from competing with domestic plants while pest control is the regulation or management of a species defined as a pest, and can be harmful for a person's health. It is an integral part of agricultural sciences. This chapter will provide an integrated understanding of weed and pest control.

Weed Control

Weed control is the botanical component of pest control, which attempts to stop weeds, especially noxious or injurious weeds, from competing with domesticated plants and livestock. Many strategies have been developed in order to contain these plants.

The original strategy was manual removal including ploughing, which can cut the roots of weeds. More recent approaches include herbicides (chemical weed killers) and reducing stocks by burning and/or pulverizing seeds.

A plant is often termed a "weed" when it has one or more of the following characteristics:

- Little or no recognized value (as in medicinal, material, nutritional or energy)
- Rapid growth and/or ease of germination
- Competitive with crops for space, light, water and nutrients

The definition of a weed is completely context-dependent. To one person, one plant may be a weed, and to another person it may be a desirable plant. In one place, a plant may be viewed as a weed, whereas in another place, the same plant may be desirable.

Introduction

Weeds compete with productive crops or pasture, ultimately converting productive land into unusable scrub. Weeds can be poisonous, distasteful, produce burrs, thorns or otherwise interfere with the use and management of desirable plants by contaminating harvests or interfering with livestock.

Weeds compete with crops for space, nutrients, water and light. Smaller, slower growing seedlings are more susceptible than those that are larger and more vigorous. Onions are one of the most vulnerable, because they are slow to germinate and produce slender, upright stems. By contrast broad beans produce large seedlings and suffer far fewer effects other than during periods of water shortage at the crucial time when the pods are filling out. Transplanted crops raised in sterile soil or potting compost gain a head start over germinating weeds.

Weeds also vary in their competitive abilities and according to conditions and season. Tall-growing vigorous weeds such as fat hen (*Chenopodium album*) can have the most pronounced effects on adjacent crops, although seedlings of fat hen that appear in late summer produce only small plants. Chickweed (*Stellaria media*), a low growing plant, can happily co-exist with a tall crop during the summer, but plants that have overwintered will grow rapidly in early spring and may swamp crops such as onions or spring greens.

The presence of weeds does not necessarily mean that they are damaging a crop, especially during the early growth stages when both weeds and crops can grow without interference. However, as growth proceeds they each begin to require greater amounts of water and nutrients. Estimates suggest that weed and crop can co-exist harmoniously for around three weeks before competition becomes significant. One study found that after competition had started, the final yield of onion bulbs was reduced at almost 4% per day.

Perennial weeds with bulbils, such as lesser celandine and oxalis, or with persistent underground stems such as couch grass (*Agropyron repens*) or creeping buttercup (*Ranunculus repens*) store reserves of food, and are thus able to grow faster and with more vigour than their annual counterparts. Some perennials such as couch grass exude allelopathic chemicals that inhibit the growth of other nearby plants.

Weeds can also host pests and diseases that can spread to cultivated crops. Charlock and Shepherd's purse may carry clubroot, eelworm can be harboured by chickweed, fat hen and shepherd's purse, while the cucumber mosaic virus, which can devastate the cucurbit family, is carried by a range of different weeds including chickweed and groundsel.

Insect pests often do not attack weeds. However pests such as cutworms may first attack weeds then move on to cultivated crops.

Some plants are considered weeds by some farmers and crops by others. Charlock, a common weed in the southeastern US, are weeds according to row crop growers, but are valued by beekeepers, who seek out places where it blooms all winter, thus providing pollen for honeybees and other pollinators. Its bloom resists all but a very hard freeze, and recovers once the freeze ends.

Weed Propagation

Seeds

Annual and biennial weeds such as chickweed, annual meadow grass, shepherd's purse, groundsel, fat hen, cleaver, speedwell and hairy bittercress propagate themselves by seeding. Many produce huge numbers of seed several times a season, some all year round. Groundsel can produce 1000 seed, and can continue right through a mild winter, whilst Scentless Mayweed produces over 30,000 seeds per plant. Not all of these will germinate at once, but over several seasons, lying dormant in the soil sometimes for years until exposed to light. Poppy seed can survive 80–100 years, dock 50 or more. There can be many thousands of seeds in a square foot or square metre of ground, thus and soil disturbance will produce a flush of fresh weed seedlings.

Subsurface/Surface

The most persistent perennials spread by underground creeping rhizomes that can regrow from a tiny fragment. These include couch grass, bindweed, ground elder, nettles, rosebay willow herb, Japanese knotweed, horsetail and bracken, as well as creeping thistle, whose tap roots can put out lateral roots. Other perennials put out runners that spread along the soil surface. As they creep they set down roots, enabling them to colonise bare ground with great rapidity. These include creeping buttercup and ground ivy. Yet another group of perennials propagate by stolons- stems that arch back into the ground to reroot. The most familiar of these is the bramble.

Methods

Weed control plans typically consist of many methods which are divided into biological, chemical, cultural, and physical/mechanical control.

Pesticide-free thermic weed control with a weed burner on a potato field in Dithmarschen

Physical/Mechanical Methods

Coverings

In domestic gardens, methods of weed control include covering an area of ground with a material that creates a hostile environment for weed growth, known as a *weed mat.*

Several layers of wet newspaper prevent light from reaching plants beneath, which kills them. Daily saturating the newspaper with water plant decomposition. After several weeks, all germinating weed seeds are dead.

In the case of black plastic, the greenhouse effect kills the plants. Although the black plastic sheet is effective at preventing weeds that it covers, it is difficult to achieve complete coverage. Eradicating persistent perennials may require the sheets to be left in place for at least two seasons.

Some plants are said to produce root exudates that suppress herbaceous weeds. Tagetes minuta is claimed to be effective against couch and ground elder, whilst a border of comfrey is also said to act as a barrier against the invasion of some weeds including couch. A 5–10 centimetres (2.0–3.9 in)} layer of wood chip mulch prevents most weeds from sprouting.

Gravel can serve as an inorganic mulch.

Irrigation is sometimes used as a weed control measure such as in the case of paddy fields to kill any plant other than the water-tolerant rice crop.

Manual Removal

Weeds are removed manually in large parts of India.

Many gardeners still remove weeds by manually pulling them out of the ground, making sure to include the roots that would otherwise allow them to resprout.

Hoeing off weed leaves and stems as soon as they appear can eventually weaken and kill perennials, although this will require persistence in the case of plants such as bindweed. Nettle infestations can be tackled by cutting back at least three times a year, repeated over a three-year period. Bramble can be dealt with in a similar way.

Tillage

Ploughing includes tilling of soil, intercultural ploughing and summer ploughing. Ploughing uproots weeds, causing them to die. In summer ploughing is done during deep summers. Summer ploughing also helps in killing pests.

Mechanical tilling can remove weeds around crop plants at various points in the growing process.

Thermal

Several thermal methods can control weeds.

Hot foam (foamstream) causes the cell walls to rupture, killing the plant. Weed burners heat up soil quickly and destroy superficial parts of the plants. Weed seeds are often heat resistant and even react with an increase of growth on dry heat.

Since the 19th century soil steam sterilization has been used to clean weeds completely from soil. Several research results confirm the high effectiveness of humid heat against weeds and its seeds.

Soil solarization in some circumstances is very effective at eliminating weeds while maintaining grass. Planted grass tends to have a higher heat/humidity tolerance than unwanted weeds.

Boiling water applied directly to the crown of weeds can also be an effective small weed killer. Larger weeds require three to four applications before being effective.

Seed Targeting

In 1998, the Australian Herbicide Resistance Initiative (AHRI), debuted. gathered fifteen scientists and technical staff members to conduct field surveys, collect seeds, test for resistance and study the biochemical and genetic mechanisms of resistance. A collaboration with DuPont led to a mandatory herbicide labeling program, in which each mode of action is clearly identified by a letter of the alphabet.

The key innovation of the AHRI approach has been to focus on weed seeds. Ryegrass seeds last only a few years in soil, so if farmers can prevent new seeds from arriving, the number of sprouts will shrink each year. Until the new approach farmers were unintentionally helping the seeds. Their combines loosen ryegrass seeds from their stalks and spread them over the fields. In the mid-1980s, a few farmers hitched covered trailers, called "chaff carts", behind their combines to catch the chaff and weed seeds. The collected material is then burned.

An alternative is to concentrate the seeds into a half-meter-wide strip called a windrow and burn the windrows after the harvest, destroying the seeds. Since 2003, windrow burning has been adopted by about 70% of farmers in Western Australia.

Yet another approach is the Harrington Seed Destructor, which is an adaptation of a coal pulverizing cage mill that uses steel bars whirling at up to 1500 rpm. It keeps all the organic material in the field and does not involve combustion, but kills 95% of seeds.

Cultural Methods

Stale Seed Bed

Another manual technique is the 'stale seed bed', which involves cultivating the soil, then leaving it fallow for a week or so. When the initial weeds sprout, the grower lightly hoes them away before planting the desired crop. However, even a freshly cleared bed is susceptible to airborne seed from elsewhere, as well as seed carried by passing animals on their fur, or from imported manure.

Buried Drip Irrigation

Buried drip irrigation involves burying drip tape in the subsurface near the planting bed, thereby limiting weeds access to water while also allowing crops to obtain moisture. It is most effective during dry periods.

Crop Rotation

Rotating crops with ones that kill weeds by choking them out, such as hemp, Mucuna pruriens, and other crops, can be a very effective method of weed control. It is a way to avoid the use of herbicides, and to gain the benefits of crop rotation.

Biological Methods

A biological weed control regiment can consist of biological control agents, bioherbicides, use of grazing animals, and protection of natural predators.

Animal Grazing

Companies using goats to control and eradicate leafy spurge, knapweed, and other toxic weeds have sprouted across the American West.

Chemical Methods

"Organic" Approaches

Weed control, circa 1930-40s

A mechanical weed control device: the diagonal weeder

Organic weed control involves anything other than applying manufactured chemicals. Typically a combination of methods are used to achieve satisfactory control.

Sulfur in some circumstances is accepted within British Soil Association standards.

Herbicides

The above described methods of weed control use no or very limited chemical inputs. They are preferred by organic gardeners or organic farmers.

However weed control can also be achieved by the use of herbicides. Selective herbicides kill certain targets while leaving the desired crop relatively unharmed. Some of these act by interfering with the growth of the weed and are often based on plant hormones. Herbicides are generally classified as follows:

Contact herbicides destroy only plant tissue that contacts the herbicide. Generally, these are the fastest-acting herbicides. They are ineffective on perennial plants that can re-grow from roots or tubers.

Systemic herbicides are foliar-applied and move through the plant where they destroy a greater amount of tissue. Glyphosate is currently the most used systemic herbicide.

Soil-borne herbicides are applied to the soil and are taken up by the roots of the target plant.

Pre-emergent herbicides are applied to the soil and prevent germination or early growth of weed seeds.

In agriculture large scale and systematic procedures are usually required, often by machines, such as large liquid herbicide 'floater' sprayers, or aerial application.

Bradley Method

Bradley Method of Bush Regeneration, which uses ecological processes to do much of the work. Perennial weeds also propagate by seeding; the airborne seed of the dandelion and the rose-bay willow herb parachute far and wide. Dandelion and dock also put down deep tap roots, which, although they do not spread underground, are able to regrow from any remaining piece left in the ground.

Hybrid

One method of maintaining the effectiveness of individual strategies is to combine them with others that work in complete different ways. Thus seed targeting has been combined with herbicides. In Australia seed management has been effectively combined with trifluralin and clethodim.

Resistance

Resistance occurs when a target adapts to circumvent a particular control strategy. It affects not only weed control, but antibiotics, insect control and other domains. In agriculture is mostly considered in reference to pesticides, but can defeat other strategies, e.g., when a target species becomes more drought tolerant via selection pressure.

Farming Practices

Herbicide resistance recently became a critical problem as many Australian sheep farmers switched to exclusively growing wheat in their pastures in the 1970s. In wheat fields, introduced varieties of ryegrass, while good for grazing sheep, are intense competitors with wheat. Ryegrasses produce so many seeds that, if left unchecked, they can completely choke a field. Herbicides provided excellent control, while reducing soil disrupting because of less need to plough. Within little more than a decade, ryegrass and other weeds began to develop resistance. Australian farmers evolved again and began diversifying their techniques.

In 1983, patches of ryegrass had become immune to Hoegrass, a family of herbicides that inhibit an enzyme called acetyl coenzyme A carboxylase.

Ryegrass populations were large, and had substantial genetic diversity, because farmers had planted many varieties. Ryegrass is cross-pollinated by wind, so genes shuffle frequently. Farmers sprayed inexpensive Hoegrass year after year, creating selection pressure, but were diluting the herbicide in order to save money, increasing plants survival. Hoegrass was mostly replaced by a group of herbicides that block acetolactate synthase, again helped by poor application practices. Ryegrass evolved a kind of "cross-resistance" that allowed it to rapidly break down a variety of herbicides. Australian farmers lost four classes of herbicides in only a few years. As of 2013 only two herbicide classes, called Photosystem II and long-chain fatty acid inhibitors, had become the last hope.

Pest Control

Pest control refers to the regulation or management of a species defined as a pest,and can be perceived to be detrimental to a person's health, the ecology or the economy. A practitioner of pest control is called an exterminator.

History

Pest control is at least as old as agriculture, as there has always been a need to keep crops free from pests. In order to maximize food production, it is advantageous to protect crops from competing species of plants, as well as from herbivores competing with humans.

The conventional approach was probably the first to be employed, since it is comparatively easy to destroy weeds by burning them or plowing them under, and to kill larger competing herbivores, such as crows and other birds eating seeds. Techniques such as crop rotation, companion planting (also known as intercropping or mixed cropping), and the selective breeding of pest-resistant cultivars have a long history.

In the UK, following concern about animal welfare, humane pest control and deterrence is gaining ground through the use of animal psychology rather than destruction. For instance, with the urban red fox which territorial behaviour is used against the animal, usually in conjunction with non-injurious chemical repellents. In rural areas of Britain, the use of firearms for pest control is quite common. Airguns are particularly popular for control of small pests such as rats, rabbits and grey squirrels, because of their lower power they can be used in more restrictive spaces such as gardens, where using a firearm would be unsafe.

Chemical pesticides date back 4,500 years, when the Sumerians used sulfur compounds as insecticides. The Rig Veda, which is about 4,000 years old, also mentions the use of poisonous plants for pest control. It was only with the industrialization and mechanization of agriculture in the 18th and 19th century, and the introduction of the insecticides pyrethrum and derris that chemical pest control became widespread. In the 20th century, the discovery of several synthetic insecticides, such as DDT, and herbicides boosted this development. Chemical pest control is still the predominant type of pest control today, although its long-term effects led to a renewed interest in traditional and biological pest control towards the end of the 20th century.

Causes

Living organisms evolve and increase their resistance to biological, chemical, physical or any other form of control. Unless the target population is completely exterminated or is rendered incapable of reproduction, the surviving population will inevitably acquire a tolerance of whatever pressures are brought to bear - this results in an evolutionary arms race.

Sign in Ilfracombe, England designed to help control seagull presence

Types of Pest Control

Use of Pest-Destroying Animals

Perhaps as far ago as 3000BC in Egypt, cats were being used to control pests of grain stores such as rodents. In 1939/40 a survey discovered that cats could keep a farm's population of rats down to a low level, but could not eliminate them completely. However, if the rats were cleared by trapping or poisoning, farm cats could stop them returning - at least from an area of 50 yards around a barn.

Ferrets were domesticated at least by 500 AD in Europe, being used as mousers. Mongooses have been introduced into homes to control rodents and snakes, probably at first by the ancient Egyptians.

Biological Pest Control

Biological pest control is the control of one through the control and management of natural predators and parasites. For example: mosquitoes are often controlled by putting *Bt* Bacillus thuringiensis ssp. *israelensis*, a bacterium that infects and kills mosquito larvae, in local water sources. The treatment has no known negative consequences on the remaining ecology and is safe for humans to drink. The point of biological pest control, or any natural pest control, is to eliminate a pest with minimal harm to the ecological balance of the environment in its present form.

Mechanical Pest Control

Mechanical pest control is the use of hands-on techniques as well as simple equipment and devices, that provides a protective barrier between plants and insects. For example: weeds can be controlled by being physically removed from the ground. This is referred to as tillage and is one of the oldest methods of weed control.

Physical Pest Control

Dog control van, Rekong Peo, Himachal Pradesh, India

Physical pest control is a method of getting rid of insects and small rodents by removing, attacking, setting up barriers that will prevent further destruction of one's plants, or forcing insect infestations to become visual.

Elimination of Breeding Grounds

Proper waste management and drainage of still water, eliminates the breeding ground of many pests.

Garbage provides food and shelter for many unwanted organisms, as well as an area where still water might collect and be used as a breeding ground by mosquitoes. Communities that have proper garbage collection and disposal, have far less of a problem with rats, cockroaches, mosquitoes, flies and other pests than those that don't.

Open air sewers are ample breeding ground for various pests as well. By building and maintaining a proper sewer system, this problem is eliminated.

Certain spectrums of LED light can "disrupt insects' breeding".

Poisoned Bait

Poisoned bait is a common method for controlling rat populations, however is not as effective when there are other food sources around, such as garbage. Poisoned meats have been used for centuries for killing off wolves, birds that were seen to threaten crops, and against other creatures. This can be a problem, since a carcass which has been poisoned will kill not only the targeted animal, but also every other animal which feeds on the carcass. Humans have also been killed by coming in contact with poisoned meat, or by eating an animal which had fed on a poisoned carcass. This tool is also used to manage several caterpillars e.g. Spodoptera litura, fruit flies, snails and slugs, crabs etc.

Field burning

Traditionally, after a sugar cane harvest, the fields are all burned, to kill off any rodents, insects or eggs that might be in the fields.

Hunting

Historically, in some European countries, when stray dogs and cats became too numerous, local populations gathered together to round up all animals that did not appear to have an owner and kill them. In some nations, teams of rat-catchers work at chasing rats from the field, and killing them with dogs and simple hand tools. Some communities have in the past employed a bounty system, where a town clerk will pay a set fee for every rat head brought in as proof of a rat killing.

Traps

A variety of mouse traps and rat traps are available for mice and rats, including snap traps, glue traps and live catch traps.

Pesticides

Rodent bait station, Chennai, India

Spraying pesticides by planes, trucks or by hand is a common method of pest control. Crop dusters commonly fly over farmland and spray pesticides to kill off pests that would threaten the crops. However, some pesticides may cause cancer and other health problems, as well as harming wildlife.

Space Fumigation

A project that involves a structure be covered or sealed airtight followed by the introduction of a penetrating, deadly gas at a killing concentration a long period of time (24-72hrs.). Although expensive, space fumigation targets all life stages of pests.

Space Treatment

A long term project involving fogging or misting type applicators. Liquid insecticide is dispersed in the atmosphere within a structure. Treatments do not require the evacuation or airtight sealing of a building, allowing most work within the building to continue but at the cost of the penetrating effects. Contact insecticides are generally used, minimizing the long lasting residual effects. On August 10, 1973, the Federal Register printed the definition of Space treatment as defined by the U.S. Environmental Protection Agency (EPA):

Residential & commercial building pest control service vehicle, Ypsilanti Township, Michigan

> **"** the dispersal of insecticides into the air by foggers, misters, aerosol devices or vapor dispensers for control of flying insects and exposed crawling insects **"**

Sterilization

Laboratory studies conducted with U-5897 (3-chloro-1,2-propanediol) were attempted in the early 1970s although these proved unsuccessful. Research into sterilization bait is ongoing.

In 2013, New York City tested sterilization traps in a $1.1 million study. The result was a 43% reduction in rat populations. The Chicago Transit Authority plans to test sterilization control in spring 2015. The sterilization method doesn't poison the rats or humans.

Destruction of Infected Plants

Forest services sometimes destroy all the trees in an area where some are infected with insects, if seen as necessary to prevent the insect species from spreading. Farms infested with certain insects, have been burned entirely, to prevent the pest from spreading elsewhere.

Natural Rodent Control

Example of House mouse infestation

Several wildlife rehabilitation organizations encourage natural form of rodent control through exclusion and predator support and preventing secondary poisoning altogether.

The United States Environmental Protection Agency agrees, noting in its Proposed Risk Mitigation Decision for Nine Rodenticides that "without habitat modification to make areas less attractive to commensal rodents, even eradication will not prevent new populations from recolonizing the habitat."

Repellents

Balsam fir oil from the tree Abies balsamea is an EPA approved non-toxic rodent repellent.

Acacia polyacantha subsp. *campylacantha* root emits chemical compounds that repel animals including crocodiles, snakes and rats.

References

- Bleasdale, J. K. A.; Salter, Peter John (1 January 1991). The Complete Know and Grow Vegetables. Oxford University Press. ISBN 978-0-19-286114-6.

- Ross, Merrill A.; Lembi, Carole A. (2008). Applied Weed Science: Including the Ecology and Management of Invasive Plants. Prentice Hall. p. 123. ISBN 978-0135028148.

- Fred Baur. Insect Management for Food Storage and Processing. American Association of Cereal Chemists. ISBN 0-913250-38-4.

- Richard Smith, W. Thomas Lanini, Mark Gaskell, Jeff Mitchell, Steven T. Koike, and Calvin Fouche (2000). "Weed Management for Organic Crops" (PDF). Division of Agriculture and Natural Resources, University of California. p. 1. Retrieved 11 December 2015.

- "Control methods". Department of Agriculture and Food, Government of Western Australia. Retrieved 11 December 2015.

- "Help WildCare Pursue Stricter Rodenticide Controls in California". wildcarebayarea.org/. Wild Care. Retrieved 28 February 2014.

- "Safer Rodenticide Products". epa.gov. USA Environment Protection Agency. March 2013. Retrieved 23 February 2014.

- WOODY, TODD (September 20, 2010). "A Crop Sprouts Without Soil or Sunshine". nytimes.com. The New York Times. Retrieved 28 February 2014.

- "Pesticides". National Institute of Health Sciences. National Institute of Environmental Health. Retrieved 5 April 2013.

2

Harmful Weeds and Pests

Noxious weeds or injurious weeds are injurious to agricultural crops, natural habitat or humans and livestock whereas an invasive species is a plant or fungus that is not native to a specific location. The following content elucidates about harmful weeds and pests, and gives a brief explanation on beneficial weeds, noxious weeds, invasive species, aphid and western corn rootworm.

Beneficial Weed

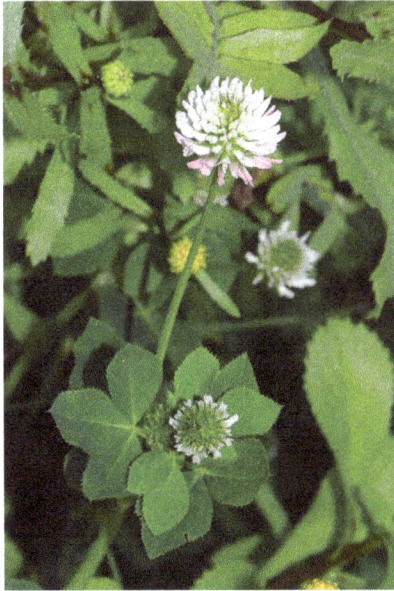

Clover was once included in grass seed mixes, because it is a legume that fertilizes the soil

A beneficial weed is a plant not generally considered domesticated and often viewed as a weed but which has some companion plant effect, is edible, contributes to soil health, or is otherwise beneficial. Beneficial weeds include many wildflowers, as well as other weeds that are commonly removed or poisoned.

Soil Health

Although erroneously assumed to compete with neighboring plants for food and moisture, some "weeds" provide the soil with nutrients, either directly or indirectly.

Dandelions benefit neighboring plant health by bringing up nutrients and moisture with its deep tap root

For example, legumes, such as white clover, if they are colonized by the right bacteria (Rhizobium most often) add nitrogen to the soil through the process of nitrogen fixation, where the bacteria has a symbiotic relationship with its hosts roots, "fixing" atmospheric nitrogen (combining it with oxygen or hydrogen) making the nitrogen plant-available ($NH4$ or $NO3$).

Others use deep tap roots to bring up nutrients and moisture from beyond the range of normal plants so that the soil improves in quality over generations of that plant's presence.

Weeds with strong, widespread roots also introduce organic matter to the earth in the form of those roots, turning hard, dense clay dirt into richer, more fertile soil.

Some plants like tomatoes and corn will "piggyback" on nearby weeds, allowing their relatively weak root systems to go deeper.

Pest Prevention

Crow garlic, like any allium, masks scents from pest insects, protecting neighboring plants

Many weeds protect nearby plants from insect pests.

Some beneficial weeds repel insects and other pests through their smell , for example alliums and wormwood. Some weeds mask a companion plant's scent, or the phero-mones of pest insects, as with ground ivy, as well as oregano and other mints.

Some also are unpleasant to small animals and ground insects, because of their spines or other features, keeping them away from an area to be protected.

Trap Crops

Some weeds act as trap crops, distracting pests away from valued plants. Insects often search for target plants by smell, and then land at random on anything green in the area of the scent. If they land on an edible "weed", they will stay there instead of going on to the intended victim. Sometimes, they actively prefer the trap crop.

Host-Finding Disruption

Recent studies on host-plant finding have shown that flying pests are far less successful if their host-plants are surrounded by any other plant or even "decoy-plants" made of green plastic, cardboard, or any other green material.

- First, they seek plants by scent. Any "weed" that has a scent reduces the odds of them finding crop plants. Examples are Crow Garlic (wild chives) and Ground Ivy (a form of wild mint), both dramatically masking both plant scent and insect pheromones. They cut down Japanese beetle infestation, and caterpillar infes-tation, for example cabbage worm, tomato hornworm, and even squash bugs.

- Second, once an insect is near its target, it avoids landing on dirt, but lands on the nearest green thing. Bare earth gardening helps them home in perfectly on the victim crop. But if one is using "green mulch", even grass or clover, the odds are that they will make what's called an "inappropriate landing" on some green thing they don't want. They will then fly a short distance at random, and land on any other green thing. If they fail to accidentally hit the right kind of plant after several tries, they give up.

- If they plan to lay eggs on the crop, weeds provide one more line of defense: Even if they find the right plant, in order to ensure that they didn't hit on a dying plant or falling leaf, they then make short leaf-to-leaf flights before laying eggs. They must land on the "right kind of leaf" enough times in sequence, before they will risk laying their eggs. The more other greenery is nearby, the harder it is for them to remain on target and get enough reinforcement. Enough "inappropri-ate landings", and they give up, heading elsewhere.

One scientific study said that simply having clover growing nearby cut the odds of cab-bage root flies hitting the right plant from 36% to 7%.

Companion Plants

Queen Anne's Lace provides shelter to nearby plants, as well as attracting predatory insects that eat pests like caterpillars, and may boost the productivity of tomato plants

Many plants can grow intercropped in the same space, because they exist on different levels in the same area, providing ground cover or working as a trellis for each other. This healthier style of horticulture is called forest gardening. Larger plants provide a wind break or shelter from noonday sun for more delicate plants.

Green Mulch

Conversely, some intercropped plants provide living mulch effect, used by inhibiting the growth of any weeds that are actually harmful, and creating a humid, cooler micro-climate around nearby plants, stabilizing soil moisture more than they consume it for themselves.

Plants such as ryegrass, red clover, and white clover are examples of "weeds" that are living mulches, often welcomed in horticulture.

Herbicide

Repel plants or fungi, through a chemical means known as allelopathy. Specific other plants can be bothered by a chemical emission through their roots or air, slowing their growth, preventing seed germination, or even killing them.

Beneficial Insects

A common companion plant benefit from many weeds is to attract and provide habitat for beneficial insects or other organisms which benefit plants.

For example, wild umbellifers attract predatory wasps and flies. The adults eat nectar, but they feed common garden pests to their offspring .

Some weeds attract ladybugs or the "good" types of nematode, or provide ground cover for predatory beetles.

Noxious Weed

noxious weed, harmful weed or injurious weed is a weed that has been designated by an agricultural authority as one that is injurious to agricultural or horticultural crops, natural habitats or ecosystems, or humans or livestock. Most noxious weeds have been introduced into an ecosystem by ignorance, mismanagement, or accident. Some noxious weeds are native. Typically they are plants that grow aggressively, multiply quickly without natural controls (native herbivores, soil chemistry, etc.), and display adverse effects through contact or ingestion. Noxious weeds are a large problem in many parts of the world, greatly affecting areas of agriculture, forest management, nature reserves, parks and other open space.

These weeds are typically agricultural pests, though many also have impacts on natural areas. Many noxious weeds have come to new regions and countries through contaminated shipments of feed and crop seeds or intentional introductions such as ornamental plants for horticultural use.

Types

There are types of noxious weeds that are harmful or poisonous to humans, domesticated grazing animals, and wildlife. Open fields and grazing pastures with disturbed soils and open sunlight are often more susceptible. Protecting grazing animals from toxic weeds in their primary feeding areas is therefore important.

Control

Some guidelines to prevent the spread of noxious weeds are:

1. Avoid driving through noxious weed-infested areas.

2. Avoid transporting or planting seeds and plants that one can't identify.

3. For noxious weeds in flower or with seeds on plants, pulling 'gently' out and placing in a secure closable bag is recommended. Disposal such as hot composting or contained burning is done when safe and practical for the specific plant. Burning poison ivy can be fatal to humans.

4. Using only certified weed-free seeds for crops or gardens.

Maintaining control of noxious weeds is important for the health of habitats, livestock, wildlife and native plants, and of humans of all ages. How to control noxious weeds depends on the surrounding environment and habitats, the weed species, the availability of equipment, labor, supplies, and financial resources. Laws often require that noxious weed control funding from governmental agencies must be used for eradication, invasion prevention, or native habitat and plant community restoration project scopes.

Noxious Weeds by Country

Australia

In Australia, the term "noxious weed" is used by state and territorial governments.

Canada

In Canada, constitutional responsibility for the regulation of agriculture and the environment are shared between the federal and provincial governments. The federal government through the Canadian Food Inspection Agency (CFIA) regulates invasive plants under the authority of the *Plant Protection Act*, the *Seeds Act* and statutory regulations. Certain plant species have been designated by the CFIA as noxious weeds in the *Weed Seeds Order*.

Each province also produces its own list of prohibited weeds. In Alberta, for example, a new *Weed Control Act* was proclaimed in 2010 with two weed designations: "prohibited noxious" (46 species) which are banned across Alberta, and "noxious" (29 species) which can be restricted at the discretion of local authorities.

New Zealand

New Zealand has had a series of Acts of Parliament relating to noxious weeds: Noxious Weeds Act, 1908, Noxious Weeds Act 1950, and the Noxious Plants Act 1978. The 1978 Act was repealed by the Biosecurity Act 1993 and words such as pest, organism and species are used in the new Act, rather than "noxious". Consequently, the term noxious weed is no longer used for official publications in New Zealand.

United Kingdom

The *Weeds Act, 1959* is described as *"Preventing the spread of harmful or injurious weeds"*, and is mainly relevant to farmers and other rural settings rather than the allotment or garden-scale grower. Five 'injurious' (that is, likely to be harmful to agricultural production) weeds are covered by the provisions of the Weeds Act. These are:

- Spear thistle (*Cirsium vulgare*)
- Creeping, or field, thistle (*Cirsium arvense*)
- Curled Dock (*Rumex crispus*)

- Broad leaved dock (*Rumex obtusifolius*)

- Common ragwort (*Senecio jacobaea*)

The Department for Environment, Food and Rural Affairs (DEFRA) provides guidance for the removal of these weeds from infested land. Much of this is oriented towards the use of herbicides.

The act does not place any automatic legal responsibility on landowners to control the weeds, but they may be ordered to control them. Most common farmland weeds are not "injurious" within the meaning of the Weeds Act and many such plant species have conservation and environmental value. DEFRA has a duty to try to achieve reasonable balance among different interests. These include agriculture, countryside conservation and the general public.

Section 14 of the *Wildlife and Countryside Act 1981*, makes it an offence to plant or grow certain specified plants in the wild, including Giant Hogweed and Japanese Knotweed. Some local authorities have bye-laws controlling these plants. There is no statutory requirement for landown-ers to remove these plants from their property.

United States

The federal government defines noxious weeds under the Federal Noxious Weed Act of 1974. Noxious weeds are also defined by the state governments in the United States.

Invasive Species

An invasive species is a plant, fungus, or animal species that is not native to a specific location (an introduced species), and which has a tendency to spread to a degree believed to cause damage to the environment, human economy or human health.

One study pointed out widely divergent perceptions of the criteria for invasive species among researchers (p. 135) and concerns with the subjectivity of the term "invasive" (p. 136). Some of the alternate usages of the term are below:

The term as most often used applies to introduced species (also called "non-indigenous" or "non-native") that adversely affect the habitats and bioregions they invade economically, environmentally, or ecologically. Such invasive species may be either plants or animals and may disrupt by dominating a region, wilderness areas, particular habitats, or wildland-urban interface land from loss of natural controls (such as predators or herbivores). This includes non-native invasive plant species labeled as exotic pest plants and invasive exotics growing in native plant communities. It has been used in this sense by government organizations as well as conservation groups such as the International Union for Conservation of Nature (IUCN) and the California Native Plant Society. The European Union defines "Invasive Alien Species" as those that are, firstly, outside their natural distribution area, and secondly, threaten biological diversity. It is also used by land managers, botanists, researchers, horticulturalists, conservationists, and the public for noxious weeds. The kudzu vine (*Pueraria lobata*), Andean Pampas grass (*Cortaderia jubata*), and yellow starthistle (*Centaurea solstitialis*) are examples.

An alternate usage broadens the term to include indigenous or "native" species along with *non-native* species, that have colonized natural areas (p. 136). Deer are an example, considered to be overpopulating their native zones and adjacent suburban gardens, by some in the Northeastern and Pacific Coast regions of the United States.

Sometimes the term is used to describe a non-native or introduced species that has become widespread (p. 136). However, not every introduced species has adverse effects on the environment. A nonadverse example is the common goldfish (Carassius auratus), which is found throughout the United States, but rarely achieves high densities (p. 136).

Causes

Scientists include species- and ecosystem factors among the mechanisms that when combined, establish invasiveness in a newly introduced species.

Species-Based Mechanisms

While all species compete to survive, invasive species appear to have specific traits or specific combinations of traits that allow them to outcompete native species. In some cases, the competition is about rates of growth and reproduction. In other cases, species interact with each other more directly.

Researchers disagree about the usefulness of traits as invasiveness markers. One study found that of a list of invasive and noninvasive species, 86% of the invasive species could be identified from the traits alone. Another study found invasive species tended to have only a small subset of the presumed traits and that many similar traits were found in noninvasive species, requiring other explanations. Common invasive species traits include the following:

- Fast growth
- Rapid reproduction
- High dispersal ability
- Phenotypic plasticity (the ability to alter growth form to suit current conditions)
- Tolerance of a wide range of environmental conditions (Ecological competence)

- Ability to live off of a wide range of food types (generalist)

- Association with humans

- Prior successful invasions

Typically, an introduced species must survive at low population densities before it becomes invasive in a new location. At low population densities, it can be difficult for the introduced species to reproduce and maintain itself in a new location, so a species might reach a location multiple times before it becomes established. Repeated patterns of human movement, such as ships sailing to and from ports or cars driving up and down highways offer repeated opportunities for establishment (also known as a high propagule pressure).

An introduced species might become invasive if it can outcompete native species for resources such as nutrients, light, physical space, water, or food. If these species evolved under great competition or predation, then the new environment may host fewer able competitors, allowing the invader to proliferate quickly. Ecosystems in which are being used to their fullest capacity by native species can be modeled as zero-sum systems in which any gain for the invader is a loss for the native. However, such unilateral competitive superiority (and extinction of native species with increased populations of the invader) is not the rule. Invasive species often coexist with native species for an extended time, and gradually, the superior competitive ability of an invasive species becomes apparent as its population grows larger and denser and it adapts to its new location.

Lantana growing in abandoned citrus plantation; Moshav Sdei Hemed, Israel

An invasive species might be able to use resources that were previously unavailable to native species, such as deep water sources accessed by a long taproot, or an ability to live on previously uninhabited soil types. For example, barbed goatgrass (Aegilops triuncialis) was introduced to California on serpentine soils, which have low water-retention, low nutrient levels, a high magnesium/calcium ratio, and possible heavy metal toxicity. Plant populations on these soils tend to show low density, but goatgrass can form dense stands on these soils and crowd out native species that have adapted poorly to serpentine soils.

Invasive species might alters its environment by releasing chemical compounds, modifying abiotic factors, or affecting the behaviour of herbivores, creating a positive or negative impact on other species. Some species, like Kalanchoe daigremontana, produce allelopathic compounds, that might have an inhibitory effect on competing species. Other species like Stapelia gigantea facilitates the recruitment of seedlings of other species in arid environments by providing appropriate microclimatic conditions and preventing herbivory in early stages of development.

Another examples are Centaurea solstitialis (yellow starthistle) and *Centaurea diffusa* (diffuse knapweed). These Eastern European noxious weeds have spread through the western and West Coast states. Experiments show that 8-hydroxyquinoline, a chemical produced at the root of *C. diffusa*, has a negative effect only on plants that have not co-evolved with it. Such co-evolved native plants have also evolved defenses. *C. diffusa* and *C. solstitialis* do not appear in their native habitats to be overwhelmingly successful competitors. Success or lack of success in one habitat does not necessarily imply success in others. Conversely, examining habitats in which a species is less successful can reveal novel weapons to defeat invasiveness.

Changes in fire regimens are another form of facilitation. Bromus tectorum, originally from Eurasia, is highly fire-adapted. It not only spreads rapidly after burning but also increases the frequency and intensity (heat) of fires by providing large amounts of dry detritus during the fire season in western North America. In areas where it is widespread, it has altered the local fire regimen so much that native plants cannot survive the frequent fires, allowing *B. tectorum* to further extend and maintain dominance in its introduced range.

Facilitation also occurs where one species physically modifies a habitat in ways that are advantageous to other species. For example, zebra mussels increase habitat complexity on lake floors, providing crevices in which invertebrates live. This increase in complexity, together with the nutrition provided by the waste products of mussel filter-feeding, increases the density and diversity of benthic invertebrate communities.

Ecosystem-Based Mechanisms

In ecosystems, the amount of available resources and the extent to which those resources are used by organisms determines the effects of additional species on the ecosystem. In stable ecosystems, equilibrium exists in the use of available resources. These mechanisms describe a situation in which the ecosystem has suffered a disturbance, which changes the fundamental nature of the ecosystem.

When changes such as a forest fire occur, normal succession favors native grasses and forbs. An introduced species that can spread faster than natives can use resources that would have been available to native species, squeezing them out. Nitrogen and phosphorus are often the limiting factors in these situations.

Every species occupies a niche in its native ecosystem; some species fill large and varied roles, while others are highly specialized. Some invading species fill niches that are not used by native species, and they also can create new niches. An example of this type can be found within the Lampropholis delicata species of skink.

Ecosystem changes can alter species' distributions. For example, edge effects describe what happens when part of an ecosystem is disturbed as when land is cleared for agriculture. The boundary between remaining undisturbed habitat and the newly cleared land itself forms a distinct habitat, creating new winners and losers and possibly hosting species that would not thrive outside the boundary habitat.

One interesting finding in studies of invasive species has shown that introduced populations have great potential for rapid adaptation and this is used to explain how so many introduced species are able to establish and become invasive in new environments. When bottlenecks and founder effects cause a great decrease in the population size, the individuals begin to show additive variance as opposed to epistatic variance. This conversion can actually lead to increased variance in the founding populations which then allows for rapid adaptive evolution. Following invasion events, selection may initially act on the capacity to disperse as well as physiological tolerance to the new stressors in the environment. Adaptation then proceeds to respond to the selective pressures of the new environment. These responses would most likely be due to temperature and climate change, or the presence of native species whether it be predator or prey. Adaptations include changes in morphology, physiology, phenology, and plasticity.

Rapid adaptive evolution in these species leads to offspring that have higher fitness and are better suited for their environment. Intraspecific phenotypic plasticity, pre-adaptation and post-introduction evolution are all major factors in adaptive evolution. Plasticity in populations allows room for changes to better suit the individual in its environment. This is key in adaptive evolution because the main goal is how to best be suited to the ecosystem that the species has been introduced. The ability to accomplish this as quickly as possible will lead to a population with a very high fitness. Pre-adaptations and evolution after the initial introduction also play a role in the success of the introduced species. If the species has adapted to a similar ecosystem or contains traits that happen to be well suited to the area that it is introduced, it is more likely to fare better in the new environment. This, in addition to evolution that takes place after introduction, all determine if the species will be able to become established in the new ecosystem and if it will reproduce and thrive.

Ecology

Traits of Invaded Ecosystems

In 1958, Charles S. Elton claimed that ecosystems with higher species diversity were less subject to invasive species because of fewer available niches. Other ecologists later pointed to highly diverse, but heavily invaded ecosystems and argued that ecosystems with high species diversity were more susceptible to invasion.

This debate hinged on the spatial scale at which invasion studies were performed, and the issue of how diversity affects susceptibility remained unresolved as of 2011. Small-scale studies tended to show a negative relationship between diversity and invasion, while large-scale studies tended to show the reverse. The latter result may be a side-effect of invasives' ability to capitalize on increased resource availability and weaker species interactions that are more common when larger samples are considered.

The brown tree snake (*Boiga irregularis*)

Invasion was more likely in ecosystems that were similar to the one in which the potential invader evolved. Island ecosystems may be more prone to invasion because their species faced few strong competitors and predators, or because their distance from colonizing species populations makes them more likely to have "open" niches. An example of this phenomenon was the decimation of native bird populations on Guam by the invasive brown tree snake. Conversely, invaded ecosystems may lack the natural competitors and predators that check invasives' growth in their native ecosystems.

Invaded ecosystems may have experienced disturbance, typically human-induced. Such a disturbance may give invasive species a chance to establish themselves with less competition from natives less able to adapt to a disturbed ecosystem.

Vectors

Non-native species have many vectors, including biogenic vectors, but most invasions are associated with human activity. Natural range extensions are common in many species, but the rate and magnitude of human-mediated extensions in these species tend to be much larger than natural extensions, and humans typically carry specimens greater distances than natural forces.

An early human vector occurred when prehistoric humans introduced the Pacific rat (*Rattus exulans*) to Polynesia.

Chinese mitten crab (*Eriocheir sinensis*)

Vectors include plants or seeds imported for horticulture. The pet trade moves animals across borders, where they can escape and become invasive. Organisms stow away on transport vehicles. Ballast water taken up at sea and released in port by transoceanic vessels is the largest vector for non-native aquatic species invasions. Around the world on the average day, more than 3,000 different species of aquatic life may be transported on these vessels. For example, freshwater zebra mussels, native to the Black, Caspian and Azov seas, probably reached the Great Lakes via ballast water from a transoceanic vessel. Although the zebra mussel invasion was first noted in 1988, and a mitigation plan was successfully implemented shortly thereafter, the plan had (and continued to have as of 2005) a serious flaw or loophole, whereby ships that are loaded with cargo when they reach the Seaway need not be tested, but all the same they transfer ballast 'puddles' between Seaway ports.

The arrival of invasive propagules to a new site is a function of the site's invasibility.

Species have also been introduced intentionally. For example, to feel more "at home," American colonists formed "Acclimation Societies" that repeatedly imported birds that were native to Europe to North America and other distant lands. In 2008, U.S. postal workers in Pennsylvania noticed noises coming from inside a box from Taiwan; the box contained more than two dozen live beetles. Agricultural Research Service entomologists identified them as rhinoceros beetle, hercules beetle, and king stag beetle. Because these species were not native to the U.S., they could have threatened native ecosystems. To prevent exotic species from becoming a problem in the U.S., special handling and permits are required when living materials are shipped from foreign countries. USDA programs such as Smuggling Interdiction and Trade Compliance (SITC) attempt to prevent exotic species outbreaks in America.

Economics plays a major role in exotic species introduction. High demand for the valuable Chinese mitten crab is one explanation for the possible intentional release of the species in foreign waters.

Impacts of Wildfire

Invasive species often exploit disturbances to an ecosystem (wildfires, roads, foot trails) to colonize an area. Large wildfires can sterilize soils, while adding a variety of nutrients. In the resulting free-for-all, formerly entrenched species lose their advantage, leaving more room for invasives. In such circumstances plants that can regenerate from their roots have an advantage. Non-natives with this ability can benefit from a low intensity fire burns that removes surface vegetation, leaving natives that rely on seeds for propagation to find their niches occupied when their seeds finally sprout.

Impact of wildfire suppression on spreading

Wildfires often occur in remote areas, needing fire suppression crews to travel through pristine forest to reach the site. The crews can bring invasive seeds with them. If any of these stowaway seeds become established, a thriving colony of invasives can erupt in as few as six weeks, after which controlling the outbreak can need years of continued attention to prevent further spread. Also, disturbing the soil surface, such as cutting firebreaks, destroys native cover, exposes soil, and can accelerate invasions. In suburban and wildland-urban interface areas, the vegetation clearance and brush removal ordinances of municipalities for defensible space can result in excessive removal of native shrubs and perennials that exposes the soil to more light and less competition for invasive plant species.

Fire suppression vehicles are often major culprits in such outbreaks, as the vehicles are often driven on back roads often overgrown with invasive plant species. The undercarriage of the vehicle becomes a prime vessel of transport. In response, on large fires, washing stations "decontaminate" vehicles before engaging in suppression activities. Large wildfires attract firefighters from remote places, further increasing the potential for seed transport.

Effects

An American alligator attacking a Burmese python in Florida; the Burmese python is an invasive species which is posing a threat to many indigenous species, including the alligator

Ecological

Land clearing and human habitation put significant pressure on local species. Disturbed habitats are prone to invasions that can have adverse effects on local ecosystems, changing ecosystem functions. A species of wetland plant known as ʻaeʻae in Hawaii (the indigenous Bacopa monnieri) is regarded as a pest species in artificially manipulated water bird refuges because it quickly covers shallow mudflats established for endangered Hawaiian stilt (Himantopus mexicanus knudseni), making these undesirable feeding areas for the birds.

Multiple successive introductions of different non-native species can have interactive effects; the introduction of a second non-native species can enable the first invasive species to flourish. Examples of this are the introductions of the amethyst gem clam (Gemma gemma) and the European green crab (Carcinus maenas). The gem clam was introduced into California's Bodega Harbor from the East Coast of the United States a century ago. It had been found in small quantities in the harbor but had never displaced the native clam species (Nutricola spp.). In the mid-1990s, the introduction of the European green crab, found to prey preferentially on the native clams, resulted in a decline of the native clams and an increase of the introduced clam populations.

In the Waterberg region of South Africa, cattle grazing over the past six centuries has allowed invasive scrub and small trees to displace much of the original grassland, resulting in a massive reduction in forage for native bovids and other grazers. Since the 1970s, large scale efforts have been underway to reduce invasive species; partial success has led to re-establishment of many species that had dwindled or left the region. Examples of these species are giraffe, blue wildebeest, impala, kudu and white rhino.

Invasive species can change the functions of ecosystems. For example, invasive plants can alter the fire regimen (cheatgrass, Bromus tectorum), nutrient cycling (smooth cordgrass Spartina alterniflora), and hydrology (Tamarix) in native ecosystems. Invasive species that are closely related to rare native species have the potential to hybridize with the native species. Harmful effects of hybridization have led to a decline and even extinction of native species. For example, hybridization with introduced cordgrass, *Spartina alterniflora*, threatens the existence of California cordgrass (Spartina foliosa) in San Francisco Bay. Invasive species cause competition for native species and because of this 400 of the 958 endangered species under the Endangered Species Act are at risk

Geomorphological

Primary geomorphological effects of invasive plants are bioconstruction and bioprotection. For example, Kudzu Pueraria montana, a vine native to Asia was widely introduced in the southeastern USA in the early 20th century to control soil erosion. While primary effects of invasive animals are bioturbation, bioerosion, and bioconstruction. For example, invasion of Chinese mitten crab Eriocheir sinensis have resulted in higher bioturbation and bioerosion rates.

Economic

Benefits

Non-native species can have benefits. Asian oysters, for example, filter water pollutants better than native oysters. They also grow faster and withstand disease better than natives. Biologists are currently considering releasing this mollusk in the Chesapeake Bay to help restore oyster stocks and remove pollution. A recent study by the Johns Hopkins School of Public Health found the Asian oyster could significantly benefit the bay's deteriorating water quality. Additionally, some species have invaded an area so long ago that they have found their own beneficial niche in the environment. For example, L. leucozonium, shown by population genetic analysis to be an invasive species in North America, has become an important pollinator of caneberry as well as cucurbit, apple trees, and blueberry bushes.

Costs

Economic costs from invasive species can be separated into direct costs through production loss in agriculture and forestry, and management costs. Estimated damage and control cost of invasive species in the U.S. alone amount to more than $138 billion annually. Economic losses can also occur through loss of recreational and tourism revenues. When economic costs of invasions are calculated as production loss and management costs, they are low because they do not consider environmental damage; if monetary values were assigned to the extinction of species, loss in biodiversity, and loss of ecosystem services, costs from impacts of invasive species would drastically increase. The following examples from different sectors of the economy demonstrate the impact of biological invasions.

Economic Opportunities

Some invasions offer potential commercial benefits. For instance, silver carp and common carp can be harvested for human food and exported to markets already familiar with the product, or processed into pet foods, or mink feed. Vegetative invasives such as water hyacinth can be turned into fuel by methane digesters.

Invasivorism

Invasive species are flora and fauna whose introduction into a habitat disrupts the native eco-system. In response, Invasivorism is a movement that explores the idea of eating invasive species in order to control, reduce, or eliminate their populations. Chefs from around the world have begun seeking out and using invasive species as alternative ingredients. Miya's of New Haven, Connecticut created the first invasive species menu in the world. Skeptics point out that once a foreign species has entrenched itself in a new place—such as the Indo-Pacific lionfish that has now virtually taken over the waters of the Western Atlantic, Caribbean and Gulf of Mexico—eradication is almost impossible. Critics argue that encouraging consumption might have the unintended effect of spreading harmful species even more widely.

A dish that features whole fried invasive lionfish at Fish Fish of Miami, Florida

Proponents of invasivorism argue that humans have the ability to eat away any species that it has an appetite for, pointing to the many animals which humans have been able to hunt to extinction - such as the Dodo bird, the Caribbean monk seal, and the Passenger pigeon. Proponents of invasivorism also point to the success that Jamaica has had in significantly decreasing the population of lionfish by encouraging the consumption of the fish.

Plant Industry

Weeds reduce yield in agriculture, though they may provide essential nutrients. Some deep-rooted weeds can "mine" nutrients from the subsoil and deposit them on the topsoil, while others provide habitat for beneficial insects or provide foods for pest species. Many weed species are accidental introductions that accompany seeds and imported plant material. Many introduced weeds in pastures compete with native forage plants, threaten young cattle (e.g., leafy spurge, Euphorbia esula) or are unpalatable because of thorns and spines (e.g., yellow starthistle). Forage loss from invasive weeds on pastures amounts to nearly US$1 billion in the U.S. alone. A decline in pollinator services and loss of fruit production has been caused by honey bees infected by the invasive varroa mite. Introduced rats (Rattus rattus and R. norvegicus) have become serious pests on farms, destroying stored grains.

Invasive plant pathogens and insect vectors for plant diseases can also suppress agricultural yields and nursery stock. Citrus greening is a bacterial disease vectored by the invasive Asian citrus psyllid (ACP). Because of the impacts of this disease on citrus crops, citrus is under quarantine and highly regulated in areas where ACP has been found.

Aquaculture

Aquaculture is a very common vector of species introductions – mainly of species with economic potential (e.g., Oreochromis niloticus)

Forestry

Poster asking campers to not move firewood around, avoiding the spread of invasive species.

The unintentional introduction of forest pest species and plant pathogens can change forest ecology and damage the timber industry. Overall, forest ecosystems in the U.S. are widely invaded by exotic pests, plants, and pathogens.

The Asian long-horned beetle (Anoplophora glabripennis) was first introduced into the U.S. in 1996, and was expected to infect and damage millions of acres of hardwood trees. As of 2005 thirty million dollars had been spent in attempts to eradicate this pest and protect millions of trees in the affected regions. The woolly adelgid has inflicted damage on old-growth spruce, fir and hemlock forests and damages the Christmas tree industry. And the chestnut blight fungus (Cryphonectria parasitica) and Dutch elm disease (Ophiostoma novo-ulmi) are two plant pathogens with serious impacts on these two species, and forest health. Garlic mustard, Alliaria petiolata, is one of the most problematic invasive plant species in eastern North American forests. The characteristics of garlic mustard are slightly different from those of the surrounding native plants, which results in a highly successful species that is altering the composition and function of the native communities it invades. When garlic mustard invades the understory of a forest, it affects the growth rate of tree seedlings, which is likely to alter forest regeneration of impact forest composition in the future.

Tourism and Recreation

Invasive species can impact outdoor recreation, such as fishing, hunting, hiking, wildlife viewing, and water-based activities. They can damage a wide array of environmental services that are important to recreation, including, but not limited to, water quality and quantity, plant and animal diversity, and species abundance. Eiswerth states, "very little research has been performed to estimate the corresponding economic losses at spatial scales such as regions, states, and watersheds." Eurasian watermilfoil (Myriophyllum spicatum) in parts of the US, fill lakes with plants complicating fishing and boating. The very loud call of the introduced common coqui depresses real estate values in affected neighborhoods of Hawaii.

Health

Encroachment of humans into previously remote ecosystems has exposed exotic diseases such as HIV to the wider population. Introduced birds (e.g. pigeons), rodents and insects (e.g. mosquito, flea, louse and tsetse fly pests) can serve as vectors and reservoirs of human afflictions. The introduced Chinese mitten crabs are carriers of Asian lung fluke. Throughout recorded history, epidemics of human diseases, such as malaria, yellow fever, typhus, and bubonic plague, spread via these vectors. A recent example of an introduced disease is the spread of the West Nile virus, which killed humans, birds, mammals, and reptiles. Waterborne disease agents, such as cholera bacteria (Vibrio cholerae), and causative agents of harmful algal blooms are often transported via ballast water. Invasive species and accompanying control efforts can have long term public health implications. For instance, pesticides applied to treat a particular pest species could pollute soil and surface water.

Biodiversity

Biotic invasion is considered one of the five top drivers for global biodiversity loss and is increasing because of tourism and globalization. This may be particularly true in inadequately regulated fresh water systems, though quarantines and ballast water rules have improved the situation.

Invasive species may drive local native species to extinction via competitive exclusion, niche displacement, or hybridisation with related native species. Therefore, besides their economic ramifications, alien invasions may result in extensive changes in the structure, composition and global distribution of the biota of sites of introduction, leading ultimately to the homogenisation of the world's fauna and flora and the loss of biodiversity. Nevertheless, it is difficult to unequivocally attribute extinctions to a species invasion, and the few scientific studies that have done so have been with animal taxa. Concern over the impacts of invasive species on biodiversity must therefore consider the actual evidence (either ecological or economic), in relation to the potential risk.

Genetic Pollution

Native species can be threatened with extinction through the process of genetic pollution. Genetic pollution is unintentional hybridization and introgression, which leads to homogenization or replacement of local genotypes as a result of either a numerical or fitness advantage of the introduced species. Genetic pollution can operate either through introduction or through habitat modification, bringing previously isolated species into contact. Hybrids resulting from rare species that interbreed with abundant species can swamp the rarer species' gene pool. This is not always apparent from morphological observations alone. Some degree of gene flow is normal, and preserves constellations of genes and genotypes. An example of this is the interbreeding of migrating coyotes with the red wolf, in areas of eastern North Carolina where the red wolf was reintroduced.

Study

Stage	Characteristic
0	Propagules residing in a donor region
I	Traveling
II	Introduced
III	Localized and numerically rare
IVa	Widespread but rare
IVb	Localized but dominant
V	Widespread and dominant

While the study of invasive species can be done within many subfields of biology, the majority of research on invasive organisms has been within the field of ecology and geography where the issue of biological invasions is especially important. Much of the study of invasive species has been influenced by Charles Elton's 1958 book *The Ecology of Invasion by Animals and Plants* which drew upon the limited amount of research done within disparate fields to create a generalized picture of biological invasions. Studies on invasive species remained sparse until the 1990s when research in the field experienced a large amount of growth which continues to this day. This research, which has largely consisted of field observational studies, has disproportionately been concerned with terrestrial plants. The rapid growth of the field has driven a need to standardize the language used to describe invasive species and events. Despite this, little standard terminology exists within the study of invasive species which itself lacks any official designation but is commonly referred to as "Invasion ecology" or more generally "Invasion biology". This lack of standard terminology is a significant problem, and has largely arisen due to the interdisciplinary nature of the field which borrows terms from numerous disciplines such as agriculture, zoology, and pathology, as well as due to studies on invasive species being commonly performed in isolation of one another.

In an attempt to avoid the ambiguous, subjective, and pejorative vocabulary that so often accompanies discussion of invasive species even in scientific papers, Colautti and MacIsaac proposed a new nomenclature system based on biogeography rather than on taxa.

By discarding taxonomy, human health, and economic factors, this model focused only on ecological factors. The model evaluated individual populations rather than entire species. It classified each population based on its success in that environment. This model applied equally to indigenous and to introduced species, and did not automatically categorize successful introductions as harmful.

Aphid

Aphids, also known as plant lice and in Britain and the Commonwealth as greenflies, blackflies, or whiteflies (not to be confused with "jumping plant lice" or true whiteflies), are small sap-sucking insects, and members of the superfamily Aphidoidea. Many species are green but other commonly occurring species may be white and wooly or black. Aphids are among the most destructive insect pests on cultivated plants in temperate regions. The damage they do to plants has made them enemies of farmers and gardeners the world over. From a zoological standpoint they are a highly successful group of organisms. Their success is due in part to the asexual reproductive capabilities of some species.

About 4,400 species are known, all included in the family Aphididae. Around 250 species are serious pests for agriculture and forestry as well as an annoyance for gardeners. They vary in length from 1 to 10 millimetres (0.04 to 0.39 in).

Natural enemies include predatory ladybugs, hoverfly larvae, parasitic wasps, aphid midge larvae, crab spiders, lacewings, and entomopathogenic fungi such as *Lecanicillium lecanii* and the Entomophthorales.

Distribution

Aphids are distributed worldwide, but are most common in temperate zones. In contrast to many taxa, aphid species diversity is much lower in the tropics than in the temperate zones. They can migrate great distances, mainly through passive dispersal by riding on winds. For example, the currant-lettuce aphid, *Nasonovia ribisnigri*, is believed to have spread from New Zealand to Tasmania in this way. Aphids have also been spread by human transportation of infested plant materials.

Taxonomy

Aphids are in the superfamily Aphidoidea in the Sternorrhyncha division of the order Hemiptera. Late 20th-century reclassification within the Hemiptera reduced the old taxon "Homoptera" to two suborders: Sternorrhyncha (e.g., aphids, whiteflies, scales, psyllids, etc.) and Auchenorrhyncha (e.g., cicadas, leafhoppers, treehoppers, planthoppers, etc.) with the suborder Heteroptera containing a large group of insects known as the true bugs. Early 21st-century reclassifications substantially rearranged the families within Aphidoidea: some old families were reduced to subfamily rank (*e.g.*, Eriosomatidae), and many old subfamilies were elevated to family rank. The most recent authoritative classifications place all extant taxa into a single large family Aphididae. Despite their names, taxonomically, the woolly conifer aphids like the pine aphid, the spruce aphid, and the balsam woolly aphid are not true aphids, but adelgids, and lack the cornicles of true aphids.

Relation to Phylloxera and Adelgids

Aphids, adelgids, and phylloxerids are very closely related, and are all within the suborder Sternorrhyncha, the plant-sucking bugs. They are either placed in the insect superfamily Aphidoidea or into the superfamily Phylloxeroidea which contains the family Adelgidae and the family Phylloxeridae.

Like aphids, phylloxera feed on the roots, leaves, and shoots of grape plants, but unlike aphids, do not produce honeydew or cornicle secretions. Phylloxera (*Daktulosphaira vitifoliae*) are insects which caused the great French wine blight that devastated European viticulture in the 19th century.

Similarly, adelgids also feed on plant phloem. Adelgids are sometimes described as aphids, but are more properly classified as aphid-like insects, because they have no cauda or cornicles.

Anatomy

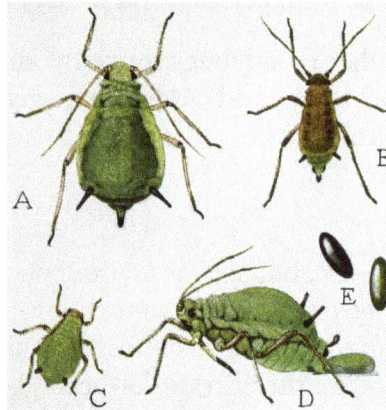

The life stages of the green apple aphid (*Aphis pomi*

Most aphids have soft bodies, which may be green, black, brown, pink, or almost colourless. Aphids have antennae with as many as six segments. They feed themselves through sucking mouthparts called stylets, enclosed in a sheath called a rostrum, which is formed from modifications of the mandible and maxilla of the insect mouthparts. They have long, thin legs and two-jointed, two-clawed tarsi.

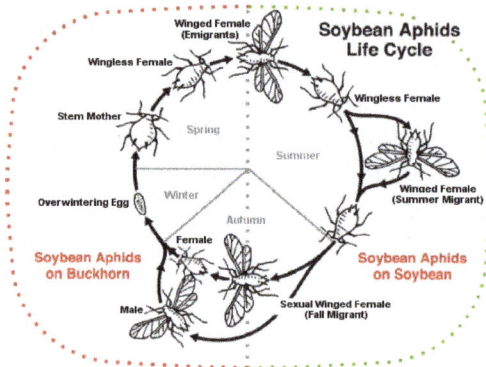

Soybean aphid lifecycle

Most aphids have a pair of cornicles (or "siphunculi"), abdominal tubes through which they exude droplets of a quick-hardening defensive fluid containing triacylglycerols, called cornicle wax. Other defensive compounds can also be produced by some types of aphids.

Aphids have a tail-like protrusion called a cauda above their rectal apertures. They have two compound eyes, and an ocular tubercle behind and above each eye, made up of three lenses (called triommatidia).

When host plant quality becomes poor or conditions become crowded, some aphid species produce winged offspring, "alates", that can disperse to other food sources. The mouthparts or eyes are smaller or missing in some species and forms.

Diet

Many aphid species are monophagous (that is, they feed on only one plant species). Others, like the green peach aphid *Myzus persicae*, feed on hundreds of plant species across many families.

Aphids passively feed on sap of phloem vessels in plants, as do many of their fellow members of Hemiptera such as scale insects and cicadas. Once a phloem vessel is punctured, the sap, which is under high pressure, is forced into the aphid's food canal. Occasionally, aphids also ingest xylem sap, which is a more dilute diet than phloem sap as the concentrations of sugars and amino acids are 1% of those in the phloem. Xylem sap is under negative hydrostatic pressure and requires active sucking, suggesting an important role in aphid physiology. As xylem sap ingestion has been observed following a dehydration period, aphids are thought to consume xylem sap to replenish their water balance; the consumption of the dilute sap of xylem permitting aphids to rehydrate. However, recent data showed aphids consume more xylem sap than expected and they notably do so when they are not dehydrated and when their fecundity decreases. This suggests aphids, and potentially, all the phloem-sap feeding species of the order Hemiptera, consume xylem sap for another reason than replenishing water balance.

Xylem sap consumption may be related to osmoregulation. High osmotic pressure in the stomach, caused by high sucrose concentration, can lead to water transfer from the hemolymph to the stomach, thus resulting in hyperosmotic stress and eventually to the death of the insect. Aphids avoid this fate by osmoregulating through several processes. Sucrose concentration is directly reduced by assimilating sucrose toward metabolism and by synthesizing oligosaccharides from several sucrose molecules, thus reducing the solute concentration and consequently the osmotic pressure. Oligasaccharides are then excreted through honeydew, explaining its high sugar concentrations, which can then be used by other animals such as ants. Furthermore, water is transferred from the hindgut, where osmotic pressure has already been reduced, to the stomach to dilute stomach content. Eventually, aphids consume xylem sap to dilute the stomach osmotic pressure. All these processes function synergetically, and enable aphids to feed on high-sucrose-concentration plant sap, as well as to adapt to varying sucrose concentrations.

Plant sap is an unbalanced diet for aphids, as it lacks essential amino acids, which aphids, like all animals, cannot synthesise, and possesses a high osmotic pressure due to its high sucrose concentration. Essential amino acids are provided to aphids by bacterial endosymbionts, harboured in special cells, bacteriocytes. These symbionts recycle glutamate, a metabolic waste of their host, into essential amino acids.

As they feed, aphids often transmit plant viruses to the plants, such as to potatoes, cereals, sugarbeets, and citrus plants. These viruses can sometimes kill the plants.

Symbioses

Ant Mutualism

Some species of ants farm aphids, protecting them on the plants they eat, eating the honey-dew the aphids release from the terminations of their alimentary canals. This is a mutualistic relationship. These dairying ants milk the aphids by stroking them with their antennae.

Ant tending aphids

Ant extracting honeydew from an aphid

Some farming ant species gather and store the aphid eggs in their nests over the winter. In the spring, the ants carry the newly hatched aphids back to the plants. Some species of dairying ants (such as the European yellow meadow ant, *Lasius flavus*) manage large herds of aphids that feed on roots of plants in the ant colony. Queens leaving to start a new colony take an aphid egg to found a new herd of underground aphids in the new colony. These farming ants protect the aphids by fighting off aphid predators.

An interesting variation in ant–aphid relationships involves lycaenid butterflies and *Myrmica* ants. For example, *Niphanda fusca* butterflies lay eggs on plants where ants tend herds of aphids. The eggs hatch as caterpillars which feed on the aphids. The ants do not defend the aphids from the caterpillars (this is due to the caterpillar producing a pheromone the ants detect making them think the caterpillar is actually one of them), but carry the caterpillars to their nest. In the nest, the ants feed the caterpillars, who in return produce honeydew for the ants. When the caterpillars reach full size, they crawl to the colony entrance and form cocoons. After two weeks, butterflies emerge and take flight. At this point the ants will attack the butterfly but the butterfly has a sticky wool like substance on their wings that disable the ants jaws meaning it can take flight without being ripped apart by the ants.

Some bees in coniferous forests also collect aphid honeydew to make forest honey.

Bacterial Endosymbiosis

Endosymbiosis with micro-organisms is common in insects, with more than 10% of insect species relying upon intracellular bacteria for their development and survival Aphids harbour a vertically transmitted (from parent to its offspring) obligate symbiosis with *Buchnera aphidicola* (Buchner) (Proteobacteria:Enterobacteriaceae), referred to as the primary symbiont, which is located inside specialised cells, the bacteriocytes. The original contamination occurred in a common ancestor 280 to 160 million years ago and has enabled aphids to exploit a new ecological niche, phloem-sap feeding on vascular plants. *B. aphidicola* provides its host with essential amino acids, which are present in low concentrations in plant sap. The stable intracellular conditions, as well as the bottleneck effect experienced during the transmission of a few bacteria from the mother to each nymph, increase the probability of transmission of mutations and gene deletions. As a result, the size of the *B. aphidicola* genome is greatly reduced, compared to its putative ancestor. Despite the apparent loss of transcription factors in the reduced genome, gene expression is highly regulated, as shown by the ten-fold variation in expression levels between different genes under normal conditions. *Buchnera aphidicola* gene transcription, although not well understood, is thought to be regulated by a small number of global transcriptional regulators and/or through nutrient supplies from the aphid host.

Some aphid colonies also harbour other bacterial symbionts, referred to as secondary symbionts due to their facultative status. They are vertically transmitted, although some studies demonstrated the possibility of horizontal transmission (from one lineage to another and possibly from one species to another). So far, the role of only some of the secondary symbionts has been described; *Regiella insecticola* plays a role in defining the host-plant range, *Hamiltonella defensa* provides resistance to parasitoids, and *Serratia symbiotica* prevents the deleterious effects of heat.

Carotenoid Synthesis

Some species of aphids have acquired the ability to synthesise red carotenoids by horizontal gene transfer from fungi. This allows otherwise green aphids to be coloured red. Other than the two-spotted spider mites aphids are the only known member of the animal kingdom with the ability to synthesise carotenoids.

Carotenoids may absorb solar energy and convert it to ATP, the first example of photoheterotrophy in animals. The carotene pigments in aphids form a layer close to the surface of the cuticle, where it is ideally placed to absorb sunlight. The excited carotenoids seem to reduce NAD to NADH which can then be oxidized in the mitochondria for energy. It is unclear why aphids should find it necessary to develop this source of energy when their diet provides them with an excess of sugars.

Reproduction

Some aphid species have unusual and complex reproductive adaptations, while others have fairly simple reproduction. Adaptations include having both sexual and asexual reproduction, creation of eggs or live nymphs, and switches between woody and herbaceous types of host plants at different times of the year.

Aphid giving birth to live young

Juvenile and adult aphids, aphid eggs, and moulting individual on *Helleborus niger*

When a sophisticated reproductive strategy is used, only females are present in the population at the beginning of the seasonal cycle (although a few species of aphids have been found to have both male and female sexes). The overwintering eggs that hatch in the spring result in females, called fundatrices. Reproduction is typically parthenogenetic and viviparous. Eggs are parthenogenetically produced without meiosis and the offspring are clonal to their mother. The embryos develop within the mothers' ovarioles, which then give live birth to first-instar female nymphs (viviparous). The offspring resemble their parents in every way except size, and are called virginoparae.

This process iterates throughout the summer, producing multiple generations that typically live 20 to 40 days. Thus, one female hatched in spring may produce thousands of descendants. For example, some species of cabbage aphids (like *Brevicoryne brassicae*) can produce up to 41 generations of females.

In autumn, aphids undergo sexual, oviparous reproduction. A change in photoperiod and temperature, or perhaps a lower food quantity or quality, causes females to parthenogenetically produce sexual females and males. The males are genetically identical to their mothers except they have one fewer sex chromosome. These sexual aphids may lack wings or even mouthparts. Sexual females and males mate, and females lay eggs that develop outside the mother. The eggs endure the winter and emerge as winged or wingless females the following spring. This is, for example, the lifecycle of the rose aphid (*Macrosiphum rosae*, or less commonly *Aphis rosae*), which may be considered typical of the family. However, in warm environments, such as in the tropics or in a greenhouse, aphids may go on reproducing asexually for many years.

Some species produce winged females in the summer, sometimes in response to low food quality or quantity. The winged females migrate to start new colonies on a new plant, often of quite a different kind. For example, the apple aphid (*Aphis pomi*), after producing many generations of wingless females on its typical food plant, gives rise to winged forms which fly away and settle on grass or corn stalks.

Some aphids have telescoping generations, that is, the parthenogenetic, viviparous female has a daughter within her, who is already parthenogenetically producing her own daughter. Thus, a female's diet can affect the body size and birth rate of more than two generations (daughters and granddaughters).

Aphid Reproduction Terminology:

Heteroecious – Host Alternating

- Egg
- Fundatrix (foundress from the first egg)
- Fundatrigeniae (daughter clones)
- Emigrant (winged female; in spring, winged aphids migrating from primary hosts infest Poaceae)
- Apterous exule (wingless female)
- Alate exule (winged female)
- Gynoparae (produce sexual females)
- Male
- Oviparae (sexual females that mate with the males)

Autoecious – Single Host

- Egg

- Fundatrix

- Apterous exule

- Alate exule

- Sexuparae (Parthenogenetic viviparous females of aphids giving rise to the sexual generation and usually developing on the secondary host, the alate forms migrating to the primary host at the end of the summer (holocyclic and heteroecious aphids).)

- Oviparae

- Males

Two adult aphids in wingless form. Pic of *Aphidoidea* taken in Belgium

Within these two host lifecycles are other forms: holocyclic (sex involved, will lead to egg production which facilitates overwintering), anholocyclic' (no sex or egg involved, reproduce parthenogenetically), and androcyclic (reproduction at end of season by parthenogenesis to produce males to contribute to holocyclic phase).

The bird cherry-oat aphid is an example of a host-alternating species (as implied by the double name), that starts its lifecycle with a large, highly fecund fundatrix. Her offspring then proceed to grow and produce emigrants which develop on the bird cherry before flying to the oat species where they continue feeding. The subsequent apterous exules feed solely on the oats and eventually lead to growth of gynoparae which will return to the bird cherry, where they will produce males and oviparae, which in turn will reproduce, giving eggs for the next year.

In heteroecious species, the aphids spend winter on tree or bush primary hosts; in summer, they migrate to their secondary host on a herbaceous plant, then the gynoparae return to the tree in autumn. The pea aphid has a primary host of a perennial vetch and secondary of the annual pea. This is likely due to the decline of food quality in trees during the summer, as well as overcrowding amongst aphids which they sense when they bump into each other too often. The heteroecious lifecycle (which is mainly linked to consumption of angiosperms and represents 10% of all aphids) is believed to have evolved from the ancestral autoecious form (on conifers); this is believed to have reverted to the ancestral form in some species that were once heteroecious.

Four types of alate (winged) aphid morphs exist, known as polymorphisms:

- Emigrants (heteroecious only) are produced on primary hosts and migrate to secondary hosts; this is once again due to quality of food decreasing and to a lesser extent overcrowding. These aphids are capable of eating both hosts.

- Alate exules are produced on secondary host if heteroecious, if autoecious will be produced on host anyway. For the alate exules the same factors apply as for emigrants EXCEPT that crowding is more important.

- Gynoparae (heteroecious only and produced on secondary host in response to longer nights and falling temperature). Nymphs can only feed on secondary hosts, and are unable to consume the primary host.

- Males are produced on secondary hosts in heteroecious and in autoecious, normal hosts. These, too, are produced in response to longer nights and decreased temperature. Of these, only 0.6% of autumn alate migrants find host plants, i.e. gynoparae.

- Reasons aphids alternate hosts:

- Nutritional optimization (right)

- Temperature tolerance – morphs adapted to part temperature

- Oviposition and rendezvous sites

- Induced host-plant defenses - plants abscise galled tissue; evidence shows that some plants selectively drop galled leaves earlier than ungalled ones.

- Increasing chance of new clones produced

    - Autoecious (increase likelihood to meet same individual)

    - Heteroecious (decreases chance of meeting self therefore mating with different clone) - better oviposition sites on trees than herbaceous plants as herbaceous plants are annual and die in winter. Problem: survival rate of autoecious vs heteroecious is similar

- Enemy escape – using the same host plants all year round increases the risk of predators discovering the aphids. However, if alates move to other hosts, they and their offspring can circumvent them for a time. One of the problems with this is the individual plants hosting the wingless aphids that have been "left behind" will have large numbers of predators which have discovered them and feed on them.

- Fundatrix specialisation – host alternation is a constraint imposed by specialized feeding requirements of the fundatrix morph as the heteroecious lifecycle is not the optimal one.

- Many host-alternating species are the biggest aphid pests:

Aphis fabae

Metopolophium dirhodum

Myzus persicae

Rhopalosiphum padi

Aphid in Baltic amber

Evolution

Aphids probably appeared around 280 million years ago, in the early Permian period. They probably fed on plants like Cordaitales or Cycadophyta. The oldest known aphid fossil is of the species *Triassoaphis cubitus* from the Triassic. The number of species was small, but increased considerably with the appearance of angiosperms 160 million years ago. Angiosperms allowed aphids to specialise. Organs like the cornicles did not appear until the Cretaceous period.

Threats

Aphids are soft-bodied, and have a wide variety of insect predators. Aphids also are often infected by bacteria, viruses, and fungi. They are affected by the weather, such as precipitation, temperature and wind.

Insects that attack aphids include predatory Coccinellidae (lady bugs or ladybirds), hoverfly larvae (Diptera: Syrphidae), parasitic wasps, aphid midge larvae, "aphid lions" (the larvae of green lacewings), and lacewings (Neuroptera: Chrysopidae), and arachnid such as crab spiders.

Predators of Aphids

Ladybird larva consuming an aphid

Fungi that attack aphids include *Neozygites fresenii*, *Entomophthora*, *Beauveria bassiana*, *Metarhizium anisopliae*, and entomopathogenic fungi such as *Lecanicillium lecanii*. Aphids brush against the microscopic spores. These spores stick to the aphid, germinate, and penetrate the aphid's skin. The fungus grows in the aphid hemolymph (i.e., the counterpart of blood for aphids). After about 3 days, the aphid dies and the fungus releases more spores into the air. Infected aphids are covered with a woolly mass that progressively grows thicker until the aphid is obscured. Often, the visible fungus is not the type of fungus that killed the aphid, but a secondary fungus.

Hoverfly larva consuming an aphid

Aphids can be easily killed by unfavourable weather, such as late spring freezes. Excessive heat kills the symbiotic bacteria that some aphids depend on, which makes the aphids infertile. Rain prevents winged aphids from dispersing, and knocks aphids off plants and thus kills them from the impact or by starvation. However, rain cannot be relied on for aphid control.

The ladybird beetle *Propylea quatuordecimpunctata* consuming an aphid

Defenses

Aphids have little protection from predators and diseases. Some species interact with plant tissues forming a gall, an abnormal swelling of plant tissue. Aphids can live inside the gall, which provides protection from predators and the elements. A number of galling aphid species are known to produce specialised "soldier" forms, sterile nymphs with defensive features which defend the gall from invasion. For example, Alexander's horned aphids are a type of soldier aphid that has a hard exoskeleton and pincer-like mouthparts. The soldiers of gall forming aphids also carry out the job of cleaning the gall. The honeydew secreted by the aphids is coated in a powdery wax to form "liquid marbles" that the soldiers roll out of the gall through small orifices. Aphids that form closed galls use the plant's vascular system for their plumbing: the inner surfaces of the galls are highly absorbent and wastes are absorbed and carried away by the plant.

Aphid excreting defensive fluid from the cornicles

Infestation of a variety of Chinese trees by Chinese sumac aphids (*Melaphis chinensis*) can create a "Chinese gall" which is valued as a commercial product. As "Galla Chinensis", Chinese galls are used in Chinese medicine to treat coughs, diarrhoea, night sweats, dysentery and to stop intestinal and uterine bleeding. Chinese galls are also an important source of tannins.

Though aphids cannot fly for most of their life cycle, they can escape predators and accidental ingestion by herbivores by dropping off the plant they are on.

Some species of aphid, known as "woolly aphids" (Eriosomatinae), excrete a "fluffy wax coating" for protection.

The cabbage aphid, *Brevicoryne brassicae*, stores and releases chemicals that produce a violent chemical reaction and strong mustard oil smell to repel predators.

It was common at one time to suggest that the cornicles were the source of the honeydew, and this was even included in the *Shorter Oxford English Dictionary* and the 2008 edition of the *World Book Encyclopedia*. In fact, honeydew secretions are produced from the anus of the aphid, while cornicles mostly produce defensive chemicals such as waxes. There also is evidence of cornicle wax attracting aphid predators in some cases. Aphids are also known to defend themselves from attack by parasitoid wasps by kicking.

Effects on Plants

Aphids living on plant host

Plants exhibiting aphid damage can have a variety of symptoms, such as decreased growth rates, mottled leaves, yellowing, stunted growth, curled leaves, browning, wilting, low yields and death. The removal of sap creates a lack of vigour in the plant, and aphid saliva is toxic to plants. Aphids frequently transmit disease-causing organisms like plant viruses to their hosts. The green peach aphid, *Myzus persicae*, is a vector for more than 110 plant viruses. Cotton aphids (*Aphis gossypii*) often infect sugarcane, papaya and peanuts with viruses. Aphids contributed to the spread of late blight (*Phytophthora infestans*) among potatoes in the Irish potato famine of the 1840s.

The cherry aphid or black cherry aphid, *Myzus cerasi*, is responsible for some leaf curl of cherry trees. This can easily be distinguished from 'leaf curl' caused by Taphrina fungus species due to the presence of aphids beneath the leaves.

In plants which produce the phytoestrogen coumestrol, such as alfalfa, damage by aphids is linked with higher concentrations of coumestrol.

Aphid with honeydew

The coating of plants with honeydew can contribute to the spread of fungi which can damage plants. Honeydew produced by aphids has been observed to reduce the effectiveness of fungicides as well.

A hypothesis that insect feeding may improve plant fitness was floated in the mid-1970s by Owen and Wiegert. It was felt that the excess honeydew would nourish soil micro-organisms, including nitrogen fixers. In a nitrogen poor environment, this could provide an advantage to an infested plant over a noninfested plant. However, this does not appear to be supported by the observational evidence.

The damage of plants, and in particular commercial crops, has resulted in large amounts of resources and efforts being spent attempting to control the activities of aphids.

A number of species of aphids of the genus *Cinara* feed on spruce and fir in North America, but do not cause noticeable injury (Rose and Lindquist 1985). Their long feeding tubes pierce the bark to take up sap from shoots, twigs, branches, stems, and roots. Aphids of most species feed in groups and are usually attended by ants, which feed on the droplets of excreted liquid. The aphids range in colour from grey to brown or black and are less than 5 mm long. All aphids overwinter in the egg stage. Eggs are blackish and are laid singly or in rows on the needles. Six generations in 1 year are not unusual in Canada, with succeeding generations often moving to new sites on the tree, including the roots, as the season progresses. The life cycle is complex. For example, adults of the intermediate summer generations consist of females only, some winged and others wingless, which produce tiny nymphs rather than eggs. Males occur only in the late fall generation, which produces the overwintering eggs.

Control

There are various insecticides that can be used to control aphids, including synthetic insecticides and plant extracts/products that are thought to be more eco-friendly. For example, Shreth *et al.* suggested use of neem and lantana products to protect plants against aphids.

For small backyard infestations, simply spraying the plants thoroughly with a strong water jet every few days may be sufficient protection for roses and other plants. An insecticidal soap solution can be an effective household remedy to control aphids and other soft-bodied arthropods. It will only kill aphids on contact and has no residual action against aphids that arrive after application. Soap spray may damage plants, especially at higher concentrations or at temperatures above 32 °C (90 °F). Some plant species are known to be sensitive to soap sprays.

Integrated pest management of various species of aphids can be achieved using biological insecticides based on fungi such as *Lecanicillium lecanii*, *Beauveria bassiana* or *Paecilomyces fumosoroseus*.

Aphids may also be controlled by the release of natural enemies, in particular lady beetles and parasitic wasps. However, since adult lady beetles tend to fly away within 48 hours after release, without laying eggs, it requires repeated application of large numbers of lady beetles to be effective. For example, one large, heavily infested rose bush may take two applications of 1500 beetles each. In reality the only cost effective situation in which mass release of natural enemies makes sense is in closed or semi closed environments such as glasshouses or polytunnels.

Boll Weevil

The boll weevil (*Anthonomus grandis*) is a beetle which feeds on cotton buds and flowers. Thought to be native to Central Mexico, it migrated into the United States from Mexico in the late 19th century and had infested all U.S. cotton-growing areas by the 1920s, devastating the industry and the people working in the American South. During the late 20th century, it became a serious pest in South America as well. Since 1978, the Boll Weevil Eradication Program in the U.S. allowed full-scale cultivation to resume in many regions.

Lifecycle

Adult weevils overwinter in well-drained areas in or near cotton fields after diapause. They emerge and enter cotton fields from early spring through midsummer, with peak emergence in late spring, and feed on immature cotton bolls. The female lays about 200 eggs over a 10- to 12-day period. The oviposition leaves wounds on the exterior of the flower bud. The eggs hatch in 3 to 5 days within the cotton squares (larger buds before flowering), feed for 8 to 10 days, and finally pupate. The pupal stage lasts another 5 to 7 days. The lifecycle from egg to adult spans about three weeks during the summer. Under optimal conditions, 8 to 10 generations per season may occur.

A female *Catolaccus grandis* wasp is attracted by a boll weevil larva.

Boll weevils begin to die at temperatures at or below −5 °C (23 °F). Research at the University of Missouri indicates they cannot survive more than an hour at −15 °C (5 °F). The insulation offered by leaf litter, crop residues, and snow may enable the beetle to survive when air temperatures drop to these levels. The boll weevil lays its eggs inside buds and ripening bolls (fruits) of the cotton plants. The adult insect has a long snout, is grayish color, and is usually less than 6 mm long.

Other limitations on boll weevil populations include extreme heat and drought. Its natural predators include fire ants, insects, spiders, birds, and a parasitic wasp, *Catolaccus grandis*. The insects sometimes emerge from diapause before cotton buds are available.

Infestation

The insect crossed the Rio Grande near Brownsville, Texas, to enter the United States from Mexico in 1892 and reached southeastern Alabama in 1909. By the mid-1920s, it had entered all cotton-growing regions in the U.S., travelling 40 to 160 miles per year. It remains the most destructive cotton pest in North America. Since the boll weevil entered the United States, it has cost U.S. cotton producers about $13 billion, and in recent times about $300 million per year.

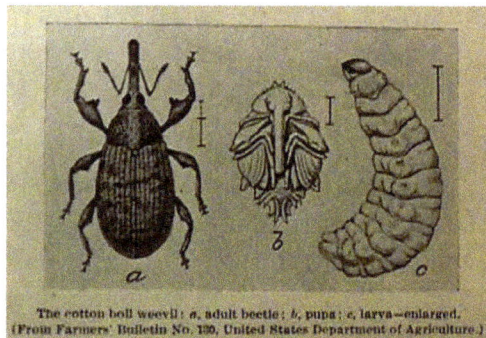

The cotton boll weevil: a, adult beetle; b, pupa; c, larva

The boll weevil contributed to the economic woes of Southern farmers during the 1920s, a situation exacerbated by the Great Depression in the 1930s.

The boll weevil appeared in Venezuela in 1949 and in Colombia in 1950. The Amazon Rainforest was thought to present a barrier to its further spread, but it was detected in Brazil in 1983, and an estimated 90% of the cotton farms in Brazil are now infested. During the 1990s, the weevil spread to Paraguay and Argentina. The International Cotton Advisory Committee has proposed a control program similar to that used in the U.S.

Control

Following World War II, the development of new pesticides such as DDT enabled U.S. farmers again to grow cotton as an economic crop. DDT was initially extremely effective, but U.S. weevil populations developed resistance by the mid-1950s. Methyl parathion, malathion, and pyrethroids were subsequently used, but environmental and resistance concerns arose as they had with DDT, and control strategies changed.

While many control methods have been investigated since the boll weevil entered the United States, insecticides have always remained the main control methods. In the 1980s, entomologists at Texas A&M University pointed to the spread of another invasive pest, the red imported fire ant, as a factor in the weevils' population decline in some areas.

Other avenues of control that have been explored include weevil-resistant strains of cotton, the parasitic wasp *Catolaccus grandis*, the fungus *Beauveria bassiana*, and the Chilo iridescent virus. Genetically engineered Bt cotton is not protected from the boll weevil.

Although it was possible to control the boll weevil, to do so was costly in terms of insecticide costs. The goal of many cotton entomologists was to eventually eradicate the pest from U. S. cotton. In 1978, a large-scale test was begun in eastern North Carolina and in Southampton County, Virginia, to determine the feasibility of eradication. Based on the success of this test, area-wide programs were begun in the 1980s to eradicate the insect from whole regions. These are based on cooperative effort by all growers together with the assistance of the Animal and Plant Health Inspection Service of the United States Department of Agriculture(USDA).

The program has been successful in eradicating boll weevils from all cotton-growing states with the exception of Texas, and most of this state is free of boll weevils. Problems along the southern border with Mexico have delayed eradication in the extreme southern portions of this state. Follow-up programs are in place in all cotton-growing states to prevent the reintroduction of the pest. These monitoring programs rely on pheromone-baited traps for detection. The boll weevil eradication program, although slow and costly, has paid off for cotton growers in reduced pesticide costs. This program and the screwworm program of the 1950s are among the biggest and most successful insect control programs in history.

Impact

The Library of Congress American Memory Project contains a number of oral history materials on the boll weevil's impact.

The boll weevil infestation has been credited with bringing about economic diversification in the Southern US, including the expansion of peanut cropping. The citizens of Enterprise, Alabama, erected the Boll Weevil Monument in 1919, perceiving that their economy had been overly dependent on cotton, and that mixed farming and manufacturing were better alternatives.

The boll weevil is the mascot for the University of Arkansas at Monticello and is listed on several "silliest" or "weirdest" mascots of all time. It was also the mascot of a short-lived minor league baseball team, the Temple Boll Weevils, which were alternatively called the "Cotton Bugs."

Colorado Potato Beetle

The Colorado potato beetle (*Leptinotarsa decemlineata*), also known as the Colorado beetle, the ten-striped spearman, the ten-lined potato beetle or the potato bug, is a major pest
of potato crops. It is approximately 10 millimetres (0.39 in) long, with a bright yellow/or-
ange body and five bold brown stripes along the length of each of its elytra. It can easily be
confused with its close cousin and look-alike, the false potato beetle.

History

The beetle was discovered in 1824 by Thomas Say from specimens collected in the Rocky
Mountains on buffalo-bur, *Solanum rostratum*. The origin of the beetle is somewhat
unclear, but it seems that Colorado and Mexico are a part of its native distribution in
southwestern North America. In about 1840, the species adopted the cultivated potato
into its host range and it rapidly became a most destructive pest of potato crops. The
large scale use of insecticides in agricultural crops effectively controlled the pest until
it became resistant to DDT in the 1950s. Other pesticides have since been used but the
insect has, over time, developed resistance to them all.

Life cycle

Colorado potato beetle larvae

Colorado potato beetle females are very prolific; they can lay as many as 800 eggs.
The eggs are yellow to orange, and are about 1 mm long. They are usually deposited
in batches of about 30 on the underside of host leaves. Development of all life stages
depends on temperature. After 4–15 days, the eggs hatch into reddish-brown larvae
with humped backs and two rows of dark brown spots on either side. They feed on the
leaves. Larvae progress through four distinct growth stages (instars). First instars are
about 1.5 mm long; the fourth is about 8 millimetres (0.31 in) long. The larvae in the
accompanying picture are third instars. The first through third instars each last about
2–3 days; the fourth, 4–7 days. Upon reaching full size, each fourth instar spends an
additional several days as a non-feeding prepupa, which can be recognized by its in-
activity and lighter coloration. The prepupae drop to the soil and burrow to a depth of

several inches, then pupate. Depending on temperature, light-regime and host quality, the adults may emerge in a few weeks to continue the life cycle, or enter diapause and delay emergence until spring. They then return to their host plant to mate and feed. In some locations, three or more generations may occur each growing season.

As a Crop Pest

Colorado beetles are a serious pest of potatoes. They may also cause significant damage to tomatoes and eggplants. Both adults and larvae feed on foliage and may skeletonize the crop. Insecticides are currently the main method of beetle control on commercial farms. However, many chemicals are often unsuccessful when used against this pest because of the beetle's ability to rapidly develop insecticide resistance. The Colorado potato beetle has developed resistance to all major insecticide classes, although not every population is resistant to every chemical. In the United Kingdom, where the Colorado beetle is a rare visitor on imported farm produce, it is a notifiable pest: any found must be reported to DEFRA.

Dutch newsreel from 1947

High fecundity usually allows Colorado potato beetle populations to withstand natural enemy pressure. Still, in the absence of insecticides natural enemies can sometimes reach densities capable of reducing Colorado potato beetle numbers below economically damaging levels. A ground beetle, *Lebia grandis*, is a predator of the eggs and larvae, and its larvae are parasitoids of the Colorado beetle's pupae. *Beauveria bassiana* (Hyphomycetes) is a pathogenic fungus that infects a wide range of insect species, including the Colorado potato beetle. It is probably the most widely used natural enemy of the Colorado potato beetle, with readily available commercial formulations that can be applied using a regular pesticide sprayer.

In Europe

In 1877, the Colorado beetle reached Germany where it was eradicated. During or immediately following WWI, it became established near USA military bases in Bordeaux and proceeded to spread by the beginning of WWII to Belgium, the Netherlands and Spain. The population increased dramatically during and immediately following WWII and spread eastward, and the beetle is now found over much of the continent. After

World War II, in the Soviet occupation zone of Germany, almost half of all potato fields were infested by the beetle by 1950. The GDR government made the claim that the beetles were dropped by American planes. In East Germany they were known as *Amikäfer* (Yankee beetles). In the EU it remains a regulated (quarantine) pest for the UK, Republic of Ireland, Balearic Islands, Cyprus, Malta and southern parts of Sweden and Finland. It is not endemic in any of these Member States, although occasional infestations occur, as in Finland in the summer of 2011, when strong winds blew from Russia, where the species is endemic.

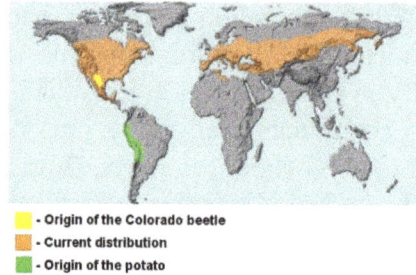

- Origin of the Colorado beetle
- Current distribution
- Origin of the potato

Native ranges of the Colorado beetle and the potato

Philately

Colorado potato beetle statue in Hédervár

The Austrian postal authority featured the beetle on a 1967 stamp. The beetle also appeared on stamps issued in Benin, Tanzania, the United Arab Emirates, and Mozambique.

The Belgian postal authority featured a drawing of the Colorado beetle and larvae on a 1934 and 1935 propaganda postcard.

In 1956, Romania issued a set of four stamps calling attention to the campaign against insect pests. The 55 Bani stamp features the Colorado potato beetle.

In Popular Culture

The beetle is documented in the documentary The Botany of Desire.

In David Javerbaum's 2011 book *The Last Testament: A Memoir by God*, the eponymous deity states that on the sixth day he tried to create the perfect beetle over 400,000 times (all failures) until he settled upon the Colorado potato beetle.

Commonly called a "tater bug", the striped thorax reminded old time musicians of the Italian style bowl-back mandolin, which are sometimes called "tater bug" mandolins.

In Czechoslovakia, a children's book called "O zlém brouku Bramborouku" ("On the evil potato-eating beetle") by Ondrej Sekora was published in 1950. It presents the Colorado beetle as an enemy for communist agriculture from the West and instructs children to fight the beetle.

In Politics

Cold War

During the Cold War the Warsaw Pact countries, fearing a food shortage, decried the beetle as a CIA plot to destroy the agriculture of the Soviet Union. Officials launched a Warsaw Pact-wide campaign to wipe out the beetle, villainizing them in propaganda posters and pulling schoolchildren from class to gather the bugs and drown them in buckets of benzene or spirit.

2014 Pro-Russian Conflict in Ukraine

Pro-Russian separatist Vostok Battalion member wearing a black and gold St. George's ribbons wristband

During the 2014 pro-Russian unrest in Ukraine, the word *kolorady*, from the Ukrainian and Russian term for Colorado beetle, (Ukrainian: жук колорадський, Russian: колорадский жук) gained popularity among Ukrainians as a derogatory term to describe pro-Russian separatists in the Donetsk and Luhansk Oblasts (provinces) of Eastern Ukraine. The nickname was a reference to the black and gold so called St. George's ribbons worn by many of the separatists as a symbol of their identity.

Phylloxera

Grape phylloxera (*Daktulosphaira vitifoliae* (Fitch 1855); family Phylloxeridae); originally described in France as *Phylloxera vastatrix*; equated to the previously described *Daktulosphaira vitifoliae, Phylloxera vitifoliae*; commonly just called phylloxera is a pest of commercial grapevines worldwide, originally native to eastern North America.

These almost microscopic, pale yellow sap-sucking insects, related to aphids, feed on the roots and leaves of grapevines (depending on the phylloxera genetic strain). On

Vitis vinifera L., the resulting deformations on roots ("nodosities" and "tuberosities") and secondary fungal infections can girdle roots, gradually cutting off the flow of nutrients and water to the vine. Nymphs also form protective galls on the undersides of grapevine leaves of some *Vitis* species and overwinter under the bark or on the vine roots; these leaf galls are typically only found on the leaves of American vines.

American vine species (such as *Vitis labrusca*) have evolved to have several natural defenses against phylloxera. The roots of the American vines exude a sticky sap that repels the nymph form when it tries to feed from the vine by clogging its mouth. If the nymph is successful in creating a feeding wound on the root, American vines respond by forming a protective layer of tissue to cover the wound and protect it from secondary bacterial or fungal infections.

Currently there is no cure for phylloxera and unlike other grape diseases such as powdery or downy mildew, there is no chemical control or response. The only successful means of controlling phylloxera has been the grafting of phylloxera resistant American rootstock (usually hybrid varieties created from the *Vitis berlandieri*, *Vitis riparia* and *Vitis rupestris* species) to more susceptible European *vinifera* vines.

Biology of Phylloxera

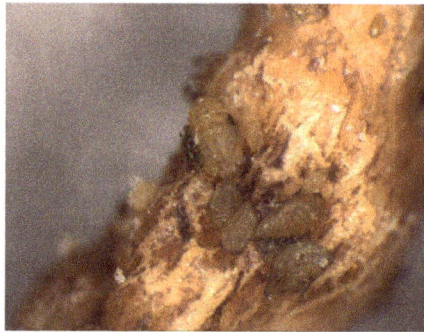

Phylloxera nymphs feeding on the roots.

The phylloxera aphid has a complex life-cycle of up to 18 stages, that can be divided into four principal forms: sexual form, leaf form, root form, and winged form.

The sexual form begins with male and female eggs laid on the underside of young grape leaves. The male and female at this stage lack a digestive system, and once hatched, they mate and then die. Before the female dies, she lays one winter egg in the bark of the vine's trunk. This egg develops into the leaf form. This nymph, the fundatrix (stem mother), climbs onto a leaf and lays eggs parthenogenetically in a leaf gall that she creates by injecting saliva into the leaf. The nymphs that hatch from these eggs may move to other leaves, or move to the roots where they begin new infections in the root form. In this form they perforate the root to find nourishment, infecting the root with a poisonous secretion that stops it from healing. This poison eventually kills the vine. This nymph reproduces by laying eggs for up to seven more generations (which also can reproduce parthenogenetically) each summer. These offspring spread to other roots of the vine, or to the roots of other vines through cracks in the soil. The generation of nymphs that hatch in the autumn hibernate in the roots and emerge next spring when the sap begins to rise. In humid areas, the nymphs develop into the winged form, else they perform the same role without wings. These nymphs start the cycle again by either staying on the vine to lay male and female eggs on the bottom side of young grape leaves, or flying to an uninfected vine to do the same.

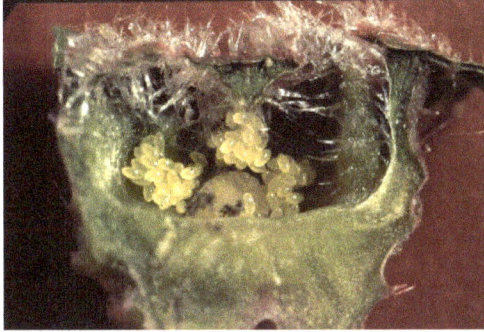

Phylloxera eggs inside a leaf gall.

Many attempts have been made to interrupt this life cycle to eradicate phylloxera, but the louse has proven to be extremely adaptable, as no one stage of the life cycle is solely dependent upon another for the propagation of the species.

Fighting the "Phylloxera Plague"

*The phylloxera, a true gourmet, finds out the best vineyards and attaches itself to the best wines*Cartoon from Punch, 6 Sep. 1890)

In the late 19th century the phylloxera epidemic destroyed most of the vineyards for wine grapes in Europe, most notably in France. Phylloxera was introduced to Europe when avid botanists in Victorian England collected specimens of American vines in the 1850s. Because phylloxera is native to North America, the native grape species are at least partially resistant. By contrast, the European wine grape *Vitis vinifera* is very susceptible to the insect. The epidemic devastated vineyards in Britain and then moved to the European mainland, destroying most of the European grape growing industry. In 1863, the first vines began to deteriorate inexplicably in the southern Rhône region of France. The problem spread rapidly across the continent. In France alone, total wine production fell from 84.5 million hectolitres in 1875 to only 23.4 million hectolitres in 1889. Some estimates hold that between two-thirds and nine-tenths of all European vineyards were destroyed.

In France, one of the desperate measures of grape growers was to bury a live toad under each vine to draw out the "poison". Areas with soils composed principally of sand or schist were spared, and the spread was slowed in dry climates, but gradually the aphid spread across the continent. A significant amount of research was devoted to finding a solution to the phylloxera problem, and two major solutions gradually emerged: grafting cuttings onto resistant rootstocks and hybridization.

Response

François Baco, creator of Baco blanc, was one of many grape breeders to introduce hybrid wine grape varieties in response to the phylloxera epidemic.

By the end of the 19th Century, hybridization became a popular avenue of research for stopping the phylloxera louse. Hybridization is the breeding of *Vitis vinifera* with resistant species. Most native American grapes are naturally phylloxera resistant (*Vitis aestivalis*, *rupestris*, and *riparia* are particularly so, while *Vitis labrusca* has a somewhat weak resistance to it) but have aromas that are off-putting to palates accustomed to European grapes. The intent of the cross was to generate a hybrid vine that was resistant to phylloxera but produced wine that did not taste like the American grape. Ironically, the hybrids tend not to be especially resistant to phylloxera, although they are much more hardy with respect to climate and other vine diseases. The new hybrid varieties have never gained the popularity of the traditional ones. In the EU they are generally banned or at least strongly discouraged from use in quality wine, although they are still in widespread use in much of North America, such as Missouri, Ontario, and upstate New York.

Grafting with Resistant Rootstock

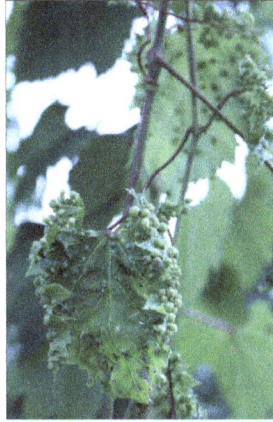

A grape leaf showing the galls that are formed during a phylloxera infestation.

Use of a resistant, or tolerant, rootstock, developed by Charles Valentine Riley in collaboration with J. E. Planchon and promoted by T. V. Munson, involved grafting a *Vitis vinifera* scion onto the roots of a resistant *Vitis aestivalis* or other American native species. This is the preferred method today, because the rootstock does not interfere with the development of the wine grapes (more technically, the genes responsible for the grapes are not in the rootstock but in the scion), and it furthermore allows the customization of the rootstock to soil and weather conditions, as well as desired vigor.

Not all rootstocks are equally resistant. Between the 1960s and the 1980s in California, many growers used a rootstock called AxR1. Even though it had already failed in many parts of the world by the early twentieth century, it was thought to be resistant by growers in California. Although phylloxera initially did not feed heavily on AxR1 roots, within twenty years, mutation and selective pressures within the phylloxera population began to overcome this rootstock, resulting in the eventual failure of most vineyards planted on AxR1. The replanting of afflicted vineyards continues today.

Many have suggested that this failure was predictable, as one parent of AxR1 is in fact a susceptible *V. vinifera* cultivar. But the transmission of phylloxera tolerance is more complex, as is demonstrated by the continued success of 41B, an F1 hybrid of *Vitis berlandieri* and *Vitis vinifera*. The full story of the planting of AxR1 in California, its recommendation, the warnings, financial consequences, and subsequent recriminations remains to be told. Modern phylloxera infestation also occurs when wineries are in need of fruit immediately, and choose to plant ungrafted vines rather than wait for grafted vines to be available.

The use of resistant American rootstock to guard against phylloxera also brought about a debate that remains unsettled to this day: whether self-rooted vines produce better wine than those that are grafted. Of course, the argument is essentially irrelevant wherever phylloxera exists. Had American rootstock not been available and used, there would be no *V. vinifera* wine industry in Europe or most places other than Chile, Washington State, and most of Australia. Cyprus was spared by the phylloxera plague, and thus its wine stock has not been grafted for phylloxera resistant purposes.

Roots that have been damaged by phylloxera.

Aftermath

The only European grapes that are natively resistant to phylloxera are the Assyrtiko grape which grows on the volcanic island of Santorini, Greece, although it is not clear whether the resistance is due to the rootstock itself or the volcanic ash on which it grows; and the Juan Garcia grape variety, autochthonous to the medieval village of Fermoselle in Spain. The Juan Garcia variety remained—untouched by phylloxera— sheltered on the vineyards planted on the man-made land terraces along the mountainous skirts on the gigantic and steep Arribes River Canyon, where the microclimatic conditions discourage the growth of phylloxera.

A grafted vine with the scion (grape variety) visible as the darker wood above the graft union and the rootstock variety below.

To escape the threat of phylloxera, wines have been produced since 1979 on the sandy beaches of Provence's Bouches-du-Rhône, which extends from the coastline of the Gard region to the waterfront village of Saintes Maries de la Mer. The sand, sun and wind in this area has been a major deterrent to phylloxera. The wine produced here is called "Vins des Sables" or "wines of the sands". In the same department, where the canal irrigation system built by the Romans still partly persists to this day, winter flooding is also practiced where possible, for instance south of the city of Tarascon. Flooding the vineyards for 50 days kills all the nymphs that overwinter in the roots or the bark at the bottom of the plant.

Some regions were so blighted by phylloxera that they never recovered, and instead the producers switched crops entirely. The island of Mallorca is one example, where almonds now substitute for vines.

Vines that Survived Phylloxera

According to wine critic and author Kerin O'Keefe, thanks to tiny parcels of vineyards throughout Europe which were inexplicably unscathed, some vineyards still exist as they were before the phylloxera devastation.

A collection of vines with grafted rootstocks.

So far, Chilean wine has remained phylloxera free. It is isolated from the world by the Atacama Desert to the north, the Pacific Ocean to the west, and the Andes Mountains to the east. Phylloxera has also never been found in several wine-growing regions of Australia, including Western Australia and South Australia. The Riesling of the Mosel region has also remained untouched by phylloxera; the parasite is unable to survive in the slate soil.

Three tiny parcels of ungrafted Pinot noir escaped phylloxera, which is used to produce Bollinger Vieilles Vignes Françaises, one of the rarest and most expensive Champagnes available.

A rare vintage port is made from ungrafted vines grown on a small parcel, called Nacional, in the heart of the Quinta do Noval estate. There is no scientific explanation as to why this plot survived while others succumbed.

Another vineyard unaffected by the phylloxera is the Lisini estate in Montalcino in Italy: a half-hectare vineyard of Sangiovese, with vines dating back to the mid-1800s.

Since 1985 the winery has produced a few bottles of Prefillossero (Italian for "before the phylloxera"). The wine has devout followers, including Italian wine critic Luigi Veronelli who inscribed on a bottle of the 1987 at the winery that drinking Prefillossero was like listening to 'the earth singing to the sky'.

Jumilla in southeastern Spain is an important area of ungrafted vineyards, mainly from Monastrell grapes. Those vineyards, however, are not immune to the louse, which is slowly advancing and destroying the fabled Pie Franco vineyard of the Casa Castillo estate, planted in 1942, i.e., when phylloxera had already been in the region for five decades.

Western Corn Rootworm

The Western corn rootworm, *Diabrotica virgifera virgifera*, is one of the most devastating corn rootworm species in North America, especially in the midwestern corn-growing areas such as Iowa. A related species, the Northern corn rootworm, *D. barberi* co-inhabits in much of the range, and is fairly similar in biology.

Two other subspecies of *D. virgifera* are described, including the Mexican corn rootworm (*Diabrotica virgifera zeae*), a significant pest in its own right, attacking corn in that country.

Corn rootworm larvae can destroy significant percentages of corn if left untreated. In the United States, current estimates show that 30,000,000 acres (12,000,000 ha) of corn (out of 80 million grown) are infested with corn rootworm. The United States Department of Agriculture estimates that corn rootworms cause $1 billion in lost revenue each year, including $800 million in yield loss and $200 million in cost of treatment for corn growers.

Life Cycle

There are many similarities in the life cycles of the northern and western corn rootworm. Both overwinter in the egg stage in the soil. Eggs, which are deposited in the soil during the summer, are American football-shaped, white, and less than 0.004 inches (0.10 mm) long. Larvae hatch in late May or early June and begin to feed on corn roots. Newly hatched larvae are small, less than .125 inches (3.2 mm) long, white worms. Corn rootworms go through three larval instars, pupate in the soil and emerge as adults in July and August. One generation emerges each year. Larvae have brown heads and a brown marking on the top of the last abdominal segment, giving them a double-headed appearance. Larvae have three pairs of legs, but these are not usually visible without magnification. After feeding for several weeks, the larvae dig a cell in the soil and molt into the pupal stage. The pupal stage is white and has the basic shape of the adult. Adult rootworms are about .25 inches (6.4 mm) long. Western corn rootworms are yellowish with a black stripe on each wing cover. Northern corn rootworm beetles are solid in color and vary from light tan to pale green.

Timing of egg hatch varies from year to year due to temperature differences and location. Males begin to emerge before females. Emergence often continues for a month or more. In years with hot, dry summers, numbers of western corn rootworm beetles may decline rapidly after mid-August, although in summers with less extreme conditions they may be found up until the first frost.

Females mate soon after emergence. Western corn rootworm females need to feed for about two weeks before they can lay eggs. Temperature and food quality influence the pre-oviposition period. Females typically lay eggs in the top 8 inches (200 mm) of soil, although they may be laid more than 12 inches (300 mm) deep, particularly if the soil surface is dry. Western corn rootworm females are more likely to lay some of their eggs below the 8-inch (200 mm) depth than northern corn rootworm females.

Feeding Habits

Rootworm larvae can complete development only on corn and a few other species of grasses. Rootworm larvae reared on other grasses (specifically, yellow foxtail) emerged as adults later and had smaller head capsule size as adults compared to larvae reared on corn. Adults feed primarily on corn silk, pollen and kernels on exposed ear tips, although they will feed on leaves and pollen of other plants. Adults begin emerging before corn reproductive tissues are present, adults may feed on leaf tissue, scraping away the green surface tissue and leaving a window-pane appearance. However, adults quickly shift to preferred green silks and pollen as they become available. Northern corn rootworm adults feed on reproductive tissues of the corn plant, but rarely feed on corn leaves. "Northern" adults are more likely than "western" adults to abandon corn and seek pollen or flowers of other plants as corn matures.

Feeding Damage

Most of the damage to corn is caused by larval feeding. Hatchlings locate roots and begin feeding on the fine root hairs, burrowing into root tips. As larvae grow, they feed on and tunnel into primary roots. When rootworms are abundant, larval feeding and deterioration of injured roots by root rot pathogens can result in roots being pruned to the stalk base. Severe root injury interferes with the roots' ability to transport water and nutrients, reduces growth and results in reduced grain production. Severe root injury may result in lodging of corn plants, making harvest more difficult. Silk feeding by adults can result in pruning at the ear tip, commonly called silk clipping. In field corn, beetle populations are occasionally high enough to cause severe silk clipping during pollen shed, which may interfere with pollination.

History of Invasions

The Western corn rootworm rapidly expanded its range in North America during the second part of the 20th century. It is now present from the southwestern region of the US Corn Belt to the east coast. It was introduced at the end of the 20th century into Europe, where it was first observed near Belgrade, Serbia in 1992. The Serbian outbreak spread north and south to include Greece to Poland and east from Italy to Ukraine. In addition to this large continuous area in Central and southeastern Europe, discontinuous outbreaks have been detected in Europe. The first was discovered near Venice, Italy, in 1998, in northwestern Italy (Piedmont) and Switzerland (canton Ticino) in 2000, northeastern Italy in 2002 (near Pordenone) and 2003 (near Udine), northern Italy (Trentino), Eastern France (Alsace), Switzerland, Belgium, the United Kingdom and the Netherlands in 2003 and the Parisian region, France in 2002, 2004 and 2005. Outbreaks detected in north Switzerland, Belgium, the Netherlands and the Parisian region did not persist. The distribution of the European corn rootworm resulted from several introductions from North America. At least three successive introductions gave rise to outbreaks detected in Serbia in 1992, the Italian Piedmont in 2000, and Ile-de-France in 2002. The European outbreaks observed in Alsace in 2003 and Ile-de-France in 2005 came from two additional introductions from North America, bringing to five the number of transatlantic introductions. The exact North American origin of the European introductions has not yet been found, but the north of the US appears to be the most likely. Small remote outbreaks in southern Germany and north-eastern Italy most likely originated from long-distance dispersal events from Central and southeastern Europe. The large European outbreak is thus likely expanding by stratified dispersal, involving both continuous diffusion and discontinuous long-distance dispersal. This latter mode of dispersal may accelerate expansion in Europe.

Management

Multiple management practices aim to control corn rootworms. These practices include corn variety selection, early planting, insecticides, crop rotation and transgenic corn varieties.

Variety

No commercial, non-transgenic resistant corn varieties are available. Several hybrid corn traits reduce damage by increasing stalk strength and root mass size. These characteristics allow a plant to better tolerate rootworm feeding, with reduced likelihood of lodging.

Early Planting

Early planted fields that have completed pollen shed are less attractive and therefore have less egg laying activity. Early fields have relatively larger root systems when rootworm feeding starts. This makes them somewhat more tolerant. Practices that promote strong root systems and a generally vigorous crop make corn more tolerant to root-worm feeding and damage.

Insecticides

Soil-applied insecticides effectively control corn rootworms. Insecticide may be warranted in areas that have a history of moderate to high damage. The number of adults present during the previous growing season is the best guide for selecting fields to be treated. However, in some areas of high insecticide use in central Nebraska, populations of corn rootworm beetles have become resistant to certain insecticides. Aldrin resistance was probably introduced independently, at least twice, from North America into Europe. Organophosphates, such as methyl-parathion, may provide effective control of both larval and adult *populations* in Central and southeastern Europe and in northwest Italy.

Crop Rotation

Crop rotation is a consistent and economical means of controlling rootworms the season following an outbreak in corn-growing areas where rootworm beetles primarily lay eggs in corn. As a way to reduce rootworm densities, it is more effective than insecticides. Corn rootworm larvae must feed on corn roots to develop and mature properly. If they hatch in a field without corn, they starve because they cannot move more than 10 to 20 inches (510 mm) in search of food. However, two rootworm biotypes survive rotation. The "soybean" variant was first discovered in central Illinois in the late 1980s and spread throughout Illinois, Indiana, southern Wisconsin and eastern Iowa. Instead of laying eggs into a corn field, the females of the soybean variant mate and then fly into a soybean field to lay their eggs, allowing the larvae to hatch in a field likely to rotate back to corn the following year. In the 1980s northern corn rootworm began to be a problem by beating the corn rotation practice with extended diapause eggs. The eggs remained in the soil for two years or more before hatching, thereby avoiding the soybean year. This adaptation has been observed in areas of northern Iowa, Minnesota and South Dakota.

Companion or second crop planting can dramatically increase rootworm populations. Corn with pumpkins or corn following pumpkins are examples of planting patterns that exacerbate rootworm feeding pressure.

Transgenics

Planting rootworm-resistant transgenic corn, is another strategy for minimizing damage. BT corn is effective at reducing root damage and is safer and often cheaper than insecticide. The transgenic traits, isolated from the common soil bacterium *Bacillus thuringiensis* strain (often referred to as Bt), produce the insect control protein.

Bt was first discovered in 1901 by Japanese biologist S. Ishiwatari as the source of disease that was killing large populations of silkworms. Bt was first used as an insecticide in 1920 and spray formulations containing either Bt bacteria or Bt proteins came into use in the 1970s for crop protection, including organic farming operations. Bt insecticides saw expanded use and development in the 1980s as an alternative to synthetic insecticides. Beginning in the 1980s, the genes responsible for making *Bt* proteins were isolated and transferred into corn plants. Bt was commercially approved in transgenic corn seed in the mid-1990s. Compared to spray formulations, transgenic plants with the Bt protein provide much more effective insect protection throughout the season. Other Bt proteins have been used to genetically modify potatoes, cotton and other types of commercial corn. The two most common brands of transgenic Bt corn are Genuity and Herculex. Genuity Smartstax combines Monsanto's VT Triple Pro, Roundup Ready 2, and Acceleron Seed Treatment System technologies, as well as Dow Chemical's Herculex Xtra and Liberty Link technologies. Acceleron, Herculex Xtra, and VT Triple Pro include traits for protection from insect damage.

Bt must be ingested to kill the insect. A susceptible larva eats the protein, which then binds to receptors in the larval gut. Binding initiates a cascade of effects that ultimately leads to death. Bt proteins are highly selective on certain categories and species of insects, eliminating insecticide use and its harmful effects to non-target organisms.

Recently, however, strains of rootworms have been discovered in several Midwest US states that exhibit Bt resistance. According to Monsanto, the "*YieldGard® VT Triple* and *Genuity® VT Triple PRO™* corn products" are affected. In 2009, four strains in Iowa were found to have field-evolved resistance to Bt corn.

By 2014 Syngenta Agrisure RW-rootworm strains had been detected in Iowa as well as glyphosate. Agrisure RW-based products entered the market in 2007. However, government officials, academics and companies lack consensus on how to define the resistance phenomenon. The affected fields constituted 0.2% of transgenic US corn acres. Further the affected areas had not been rotated with other crops.

References

- Elton, C.S. (2000) [1958]. The Ecology of Invasions by Animals and Plants. Foreword by Daniel Simberloff. Chicago: University of Chicago Press. p. 196. ISBN 0-226-20638-6.

- Ross Piper (2007). Extraordinary Animals: An Encyclopedia of Curious and Unusual Animals. Greenwood Press. pp. 6–9. ISBN 978-0-313-33922-6.

- R. L. Blackman; V. F. Eastrop (1994). Aphids on the World's Trees. An Identification and Information Guide. Wallingford: CAB International. ISBN 0-85198-877-6.

- Henry G. Stroyan (1997). "Aphid". McGraw-Hill Encyclopedia of Science and Technology (8th ed.). ISBN 0-07-911504-7.

- D. B. Fisher (2000). "Long distance transport". In Bob B. Buchanan; Wilhelm Gruissem; Russell L. Jones. Biochemistry and Molecular Biology of Plants (4th ed.). Rockville, Maryland: American

Society of Plant Physiologists. pp. 730–784. ISBN 978-0-943088-39-6.

- Timothy D. Schowalter (31 May 2011). Insect Ecology: An Ecosystem Approach. Academic Press. p. 482. ISBN 978-0-12-381351-0. Retrieved 8 November 2011.

- Herrmann, Bernd (2009). "Schauplätze und Themen der Umweltgeschichte : Umwelthistorische Miszellen aus dem Graduiertenkolleg". ISBN 9783941875234.

- Wine & Spirits Education Trust "Wine and Spirits: Understanding Wine Quality" pgs 2-5, Second Revised Edition (2012), London, ISBN 9781905819157

- "Poison Ivy Identification and Control" (PDF). Archived (PDF) from the original on 2015-06-22. Retrieved 2015-06-22.

- Lockwood, Julie L.; Martha F. Hoopes; Michael P. Marchetti (2007). Invasion Ecology (PDF). Blackwell Publishing. p. 7. Retrieved 21 January 2014.

- Sindelar, Daisy. "What's Orange and Black and Bugging Ukraine?". Radio Free Europe / Radio Liberty. Retrieved 18 May 2014.

- F. Moretzsohn; J.A. Sánchez Chávez; J.W. Tunnell, Jr. (eds.). "Invasive Species". GulfBase: Resource Database for Gulf of Mexico Research. Harte Research Institute for Gulf of Mexico Studies at Texas A&M University-Corpus Christi. Retrieved March 19, 2013.

- "National weeds lists". Department of Sustainability, Environment, Water, Population and Communities. 14 August 2012. Retrieved 25 August 2012.

- Anon. "Companion Planting for Vegetables & Plants". Country living and farm lifestyles. country-farm-lifestyles.com. Retrieved 2011-03-07.

- "Communication From The Commission To The Council, The European Parliament, The European Economic And Social Committee And The Committee Of The Regions Towards An EU Strategy On InvasFpollive Species" (PDF). Retrieved 2011-05-17.

3

Weed Control Methods

Mechanical weed control technique manages to remove weed physically by injuring or making the growing conditions unfavorable. This chapter discusses the methods of weed control in a critical manner providing key analysis to the subject matter. It elucidates methods such as mechanical weed control, bush regeneration, soil steam sterilization and irrigation. This chapter is a compilation of the various branches of weed control that form an integral part of the broader subject matter.

Mechanical Weed Control

Mechanical weed control is any physical activity that inhibits unwanted plant growth. Mechanical, or manual, weed control techniques manage weed populations through physical methods that remove, injure, kill, or make the growing conditions unfavorable. Some of these methods cause direct damage to the weeds through complete removal or causing a lethal injury. Other techniques may alter the growing environment by eliminating light, increasing the temperature of the soil, or depriving the plant of carbon dioxide or oxygen. Mechanical control techniques can be either selective or non-selective. A selective method has very little impact on non-target plants where as a non-selective method affects the entire area that is being treated. If mechanical control methods are applied at the optimal time and intensity, some weed species may be controlled or even eradicated.

Mechanical Control Methods

Weed Pulling

Pulling methods uproot and remove the weed from the soil. Weed pulling can be used to control some shrubs, tree saplings, and herbaceous plants. Annuals and tap-rooted weeds tend to be very susceptible to pulling. Many species are able to re-sprout from root segments that are left in the soil. Therefore, the effectiveness of this method is dependent on the removal of as much of the root system as possible. Well established perennial weeds are much less effectively controlled because of the difficulty of removing all of the root system and perennating plant parts. Small herbaceous weeds may be pulled by hand but larger plants may require the use of puller tools like the Weed Wrench or the Root Talon. This technique has a little to no impact on neighboring, non-target plants and has a minimal effect on the growing environment. However, pulling is labor-intensive and time consuming making it a more suitable method to use for small weed infestations.

Mowing

Mowing methods cut or shred the above ground of the weed and can prevent and reduce seed populations as well as restrict the growth of weeds. Mowing can be a very successful control method for many annual weeds. Mowing is the most effective when it is performed before the weeds are able to set seed because it can reduce the number of flower stalks and prevent the spread of more seed. However, the biology of the weed must be considered before mowing. Some weed species may sprout with increased vigor after being mowed. Also, some species are able to re-sprout from stem or root segments that are left behind after mowing. Brush cutting and weed eating are also mowing techniques that reduce the biomass of the weeds. Repeatedly removing biomass causes reduced vigor in many weed species. This method is usually used in combination with other control methods such as burning or herbicide treatments.

Mulching

Mulch is a layer of material that is spread on the ground. Compared with some other methods of weed control, mulch is relatively simple and inexpensive. Mulching smothers the weeds by excluding light and providing a physical barrier to impede their emergence. Mulching is successful with most annual weeds, however, some perennial weeds are not affected. Mulches may be organic or synthetic. Organic mulches consist of plant by products such as: pine straw, wood chips, green waste, compost, leaves, and grass clippings. Synthetic mulches, also known as ground cover fabric, can be made from materials like polyethylene, polypropylene, or polyester. The effectiveness of mulching is mostly dependent on the material used. Organic and synthetic mulches may be used in combination with each other to increase the amount of weeds controlled.

Tillage

Tillage, also known as cultivation, is the turning over of the soil. This method is more often used in agricultural crops. Tillage can be performed on a small scale with tools such as small, hand pushed rotary tillers or on a large scale with tractor mounted plows. Tillage is able to control weeds because when the soil is overturned, the vegetative parts of the plants are damaged and the root systems are exposed causing desiccation. Generally, the younger the weed is, the more readily it can be controlled with tillage. To control mature perennial weeds, repeated tillage is necessary. By continually destroying new growth and damaging the root system, the weed's food stores are depleted until it can no longer re-sprout. Also, when the soil is overturned, the soil seed bank is disrupted which can cause dormant weed seeds to germinate in the absence of the previous competitors. These new weeds can also be controlled by continued tillage until the soil seed bank is depleted.

Soil Solarization

Soil solarization is a simple method of weed control that is accomplished by covering the soil with a layer of clear or black plastic. The plastic that is covering the ground traps heat energy from the sun and raises the temperature of the soil. Many weed seeds and vegetative propagules are not able to withstand the temperatures and are killed. For this method to be most effective, it should be implemented during the summer months and the soil should be moist. Also, cool season weeds are more susceptible to soil solarization than are warm season weeds. Using black plastic as a cover excludes light which can help to control plants that are growing whereas clear plastic has been shown to produce higher soil temperatures.

Fire

Burning and flaming can be economical and practical methods of weed control if used carefully. For most plants, fire causes the cell walls to rupture when they reach a temperature of 45 °C to 55 °C. Burning is commonly used to control weeds in forests, ditches, and roadsides. Burning can be used to remove accumulated vegetation by destroying the dry, matured plant matter as well as killing the green new growth. Buried weed seeds and plant propagules may also be destroyed during burning, however, dry seeds are much less susceptible to the increased temperature. Flaming is used on a smaller scale and includes the use of a propane torch with a fan tip. Flaming may be used to control weeds along fences and paved areas or places where the soil may be too wet to hoe, dig, or till. Flaming is most effective on young weeds that are less than two inches tall but repeated treatments may control tougher perennial weeds.

Flooding

Flooding is a method of control that requires the area being treated to be saturated at a depth of 15 to 30 cm for a period of 3 to 8 weeks. The saturation of the soil reduces the availability of oxygen to the plant roots thereby killing the weed. This method has been shown to be highly effective in controlling establish perennial weeds and may also suppress annual weeds by reducing the weed seed populations.

Effects of Mechanical Control on the Environment

Mechanical methods of weed control cause physical changes in the immediate environment that may cause positive or negative effects. The suppression of the targeted weeds will open niches in the environment and may also stimulate the growth of other weeds by decreasing their competition and making their environment more favorable. If the niches are not filled by a desirable plant, they will eventually be taken over by another weed. These weed control methods also effect the structure of the soil. The use of mulches can help decrease erosion, decrease water evaporation from the soil, as well as improve the soil structure by increasing the amount of organic matter. Tillage practices can help decrease compaction and aerate the soil. On the other hand, tillage has also been shown to decrease soil moisture, increase soil erosion and runoff, as well as decrease soil microbial populations. Solarization can cause changes in the biological, physical, and chemical properties of the soil. This can cause the soil to be an unfavorable environment for native species which may be beneficial or harmful.

Bush Regeneration

Bush regeneration, a form of natural area restoration, is the term used in Australia for the ecological restoration of remnant vegetation areas, such as through the minimisation of negative disturbances, both exogenous such as exotic weeds and endogenous such as erosion. It may also attempt to recreate conditions of pre-European arrival, for example by simulating endogenous disturbances such as fire. Bush regeneration attempts to protect and enhance the floral biodiversity in an area by providing conditions conducive to the recruitment and survival of native plants. Bushcare's Major Day Out is an Australian national day of community participation in the care of bushland. In 2012 nearly 100 bushcare sites participated in this annual event. For more information go to www.bushcaresmajordayout.org.

History

Bradley Method

In the early 1960s Joan and Eileen Bradley developed a series of weed control techniques through a process of trial and error. Their work was the beginning of minimal disturbance bush regeneration in New South Wales. The Bradley method urges a naturalistic approach by encouraging the native vegetation to self-reestablish. The Bradleys used their method to successfully clear weeds from a 16 hectares (40 acres) reserve in Ashton Park, part of Sydney Harbour National Park, NSW. The process demonstrated that, following a period of consecutive 'follow up' treatments of diminishing time requirement, subsequent maintenance was needed only once or twice a year, mainly in vulnerable spots such as creek banks, roadsides, and clearings, to be maintained weed-free.

The aim of their work was to clear small niches adjacent to healthy native vegetation such that the each area will regenerate from in-situ soil seed banks or be re-colonised and stabilized by the regeneration of native plants, replacing an area previously occupied by weeds. The Bradley method follows three main principles,

- secure the best areas first

- minimise disturbance to the natural conditions (e.g. minimise soil disturbance and off-target damage).

- don't overclear, let the regenerative ability of the bush set the pace of clearance (Bradley 1988).

The priority securing of the best quality vegetation aids in preserving areas of top biodiversity which provide regeneration potential to expand these areas and reclaim areas as bushland.

Modern Practice

The adoption of minimal disturbance bush regeneration increased in the decades that followed the work of the Bradleys. Their principles have guided bushcare programs in Australia, although the inclusion of herbicide in modern bush regeneration is a notable deviation from the ideals of the Bradley sisters. In addition, rather than 'minimal disturbance', a more favoured and ecologically sound trend since the 1990s has been towards more 'appropriate disturbance' as many Australian plant communities require some level of perturbation to trigger germination from long-buried seed banks. This has led to a range of additional disturbance-based techniques (such as burns and soil disturbance) being included in the regenerator's 'tool kit' in dry forest and grassland areas. Field experience has found that, even in rainforest areas, a resilience to disturbance is evident, enabling regenerators to clear weed in a fairly extensive manner to trigger rainforest recovery. This is borne out by a thriving rainforest regeneration industry in northern NSW Australia, modelled on the pioneering work of John Stockard at Wingham Brush (Stockard 1991, Stockard 1999). The rule of thumb in all cases is to constrain clearing to that area that matches the project's follow up resources.

The increased awareness and consideration of Australia's biodiversity by citizens has incrementally increased pressure on local governments to adopt conservation programs for remnant vegetation on council land. Most peri urban councils now have some involvement in bush regeneration, either through planning, land management, volunteer support or through employment of bush regeneration practitioners. In NSW the level of coordination of bush regeneration programs through local governments is high, although in some other areas at present a lack of coordination is a serious concern in bush regeneration on public land, with only 40% of councils liaising with other councils. In such areas there may be a need for strategic management at a regional scale through Natural Resource Management Boards or non government organisations such as Trees For Life, which are involved in bushcare programs across wider areas.

Purposes

The aim of bush regeneration, also known as 'natural area restoration', is to restore and maintain ecosystem health by facilitating the natural regeneration of indigenous flora, this is usually achieved by selectively reducing the competitive interaction with invasive species, or mitigation of negative influences such as weeds or erosion.

Invasive plant species are often the greatest threat to remnant vegetation, and therefore bush regeneration is closely associated with weed abatement activities. Weed management as one aim of bush regeneration, is used to increase native plant recruitment. The management of factors such as fire and herbivory can be just as important, depending on the ecosystem being restored. In recent years research and on-ground management has begun to recognize the importance of ecosystem processes rather ecosystem composition and structure and research into other ways of facilitating native plant recruitment is increasing.

Technique

The original Bradley method of bush regeneration focuses on facilitating native plant recruitment from the seedbank, rather than planting seedlings or sowing seeds, as follows:

"Weeding a little at a time from the bush towards the weeds takes the pressure off the natives under favourable conditions. Native seeds and spores are ready in the ground and the natural environment favours plants that have evolved in it. The balance is tipped back towards regeneration. Keep it that way, by always working where the strongest area of bush meets the weakest weeds"

Currently the term 'bush regeneration' includes activities other than weed removal, such as replanting and introducing species into an area where soil, water, or fire regimes have shifted the type of plant appropriate to the area (e.g. a stormwater drain).

Weed species can be important habitat for native fauna (e.g. blackberry is important habitat for wrens and the southern brown bandicoot) and this should be taken into consideration with bush regeneration, for example by not clearing invasive species until adequate habitat alternatives have been established nearby with native vegetation.

Problems can occur when insufficient follow-up is conducted as the success of bush regeneration is dependent on allowing the native vegetation to regenerate in the area where weeds have been removed.

Soil Steam Sterilization

Soil steam sterilization (soil steaming) is a farming technique that sterilizes soil with steam in open fields or greenhouses. Pests of plant cultures such as weeds, bacteria, fungi and viruses are killed through induced hot steam which causes their cell structure to physically degenerate. Biologically, the method is considered a partial disinfection. Important heat-resistant, spore-forming bacteria survive and revitalize the soil after cooling down. Soil fatigue can be cured through the release of nutritive substances blocked within the soil. Steaming leads to a better starting position, quicker growth and strengthened resistance against plant disease and pests. Today, the application of hot steam is considered the best and most effective way to disinfect sick soil, potting soil and compost. It is being used as an alternative to bromomethane, whose production and use was curtailed by the Montreal Protocol. "Steam effectively kills pathogens by heating the soil to levels that cause protein coagulation or enzyme inactivation."

Benefits of Soil Steaming

Soil sterilization provides secure and quick relief of soils from substances and organisms harmful to plants such as:

- Bacteria
- Viruses
- Fungi
- Nematodes and
- Other Pests

Further positive effects are:

- All weed and weed seeds are killed
- Significant increase of crop yields
- Relief from soil fatigue through activation of chemical – biological reactions
- Blocked nutritive substances in the soil are tapped and made available for plants
- Alternative to Methyl Bromide and other critical chemicals in agriculture

Steaming with Superheated Steam

Through modern steaming methods with superheated steam at 180–200 °C, an optimal soil disinfection can be achieved. Soil only absorbs a small amount of humidity. Micro organisms become active once the soil has cooled down. This creates an optimal environment for instant tillage with seedlings and seeds. Additionally the method of integrated steaming can promote a target-oriented resettlement of steamed soil with beneficial organisms. In the process, the soil is first freed from all organisms and then revitalized and microbiologically buffered through the injection of a soil activator based on compost which contains a natural mixture of favorable microorganisms (e.g. Bacillus subtilis, etc.).

Different types of such steam application are also available in practice, including substrate steaming and surface steaming.

Surface Steaming

Several methods for surface steaming are in use amongst which are: area sheet steaming, the steaming hood, the steaming harrow, the steaming plough and vacuum steaming with drainage pipes or mobile pipe systems.

In order to pick the most suitable steaming method, certain factors have to be considered such as soil structure, plant culture and area performance. At present, more advanced methods are being developed, such as sandwich steaming or partially integrated sandwich steaming in order to minimize energy and cost as much as possible.

Sheet Steaming

Sheet steaming with a MSD/moeschle steam boiler (left side)

Large area sheet steaming in greenhouses using a steam injector.

Surface steaming with special sheets (sheet steaming) is a method which has been established for decades in order to steam large areas reaching from 15 to 400 m² in one step. If properly applied, sheet steaming is simple and highly economic. The usage of heat resistant, non-decomposing insulation fleece saves up to 50% energy, reduces the steaming time significantly and improves penetration. Single working step areas up to 400 m² can be steamed in 4–5 hours down to 25–30 cm depth / 90°C. The usage of heat resistant and non-decomposing synthetic insulation fleece, 5 mm thick, 500 gr / m², can reduce steaming time by about 30%. Through a steam injector or a perforated pipe, steam is injected underneath the sheet after it has been laid out and weighted with sand sacks.

The area performance in one working step depends on the capacity of the steam generator (e.g. steam boiler):

Steam capacity kg/h:	100	250	300	400	550	800	1000	1350	2000
Area m²:	15-20	30-50	50-65	60-90	80-120	130-180	180-220	220-270	300-400

The steaming time depends on soil structure as well as outside temperature and amounts to 1-1.5 hours per 10 cm steaming depth. Hereby the soil reaches a temperature of about 85°C. Milling for soil loosening is not recommended since soil structure may become too fine which reduces its penetrability for steam. The usage of spading machines is ideal for soil loosening. The best results can be achieved if the soil is cloddy at greater depth and granulated at lesser depth.

In practice, working with at least two sheets simultaneously has proven to be highly effective. While one sheet is used for steaming the other one is prepared for steam injection, therefore unnecessary steaming recesses are avoided.

Depth Steaming with Vacuum

Steaming with vacuum which is induced through a mobile or fixed installed pipe system in the depth of the area to be steamed, is the method that reaches the best penetration. Despite high capital cost, the fixed installation of drainage systems is reasonable for intensively used areas since steaming depths of up to 80 cm can be achieved.

In contrast to fixed installed drainage systems, pipes in mobile suction systems are on the surface. A central suction pipeline consisting of zinc-coated, fast-coupling pipes are connected in a regular spacing of 1.50 m and the ends of the hoses are pushed into the soil to the desired depth with a special tool.

The steaming area is covered with a special steaming sheet and weighted all around as with sheet steaming. The steam is injected underneath the sheet through an injector and protection tunnel. While with short areas up to 30 m length steam is frontally injected, with longer areas steam is induced in the middle of the beet using a T-connection branching out to both sides. As soon as the sheet is inflated to approximately 1m by the steam pressure, the suction turbine is switched on. First, the air in the soil is removed via the suction hoses. A vacuum is formed and the steam is pulled downward.

During the final phase, when the required steaming depth has been reached, the ventilator runs non-stop and surplus steam is blown out. To ensure that this surplus steam is not lost, it is fed back under the sheet.

As with all other steaming systems, a post-steaming period of approximately 20–30 minutes is required. Steaming time is approximately 1 hour per 10 cm steaming depth. The steam requirement is approximately 7–8 kg/m².

The most important requirement, as with all steaming systems, is that the soil is well loosened before steaming, to ensure optimal penetration.

Negative Pressure Technique

"Negative Pressure technique generates appropriate soil temperature at a 60 cm depth and complete control of nematodes, fungi and weeds is achieved. In this technique, the steam is introduced under the steaming sheath and forced to enter the soil profile by a negative pressure. The negative pressure is created by a fan that sucks the air out of the soil through buried perforated polypropylene pipes. This system requires a permanent installation of perforated pipes into the soil, at a depth of at least 60 cm to be protected from plough."

Steaming with Hoods

Half automatic steaming hood with three wings in greenhouse

A steaming hood is a mobile device consisting of corrosion-resistant materials such as aluminum, which is put down onto the area to be steamed. In contrast to sheet steaming, cost-intensive working steps such as laying out and weighting the sheets don't occur, however the area steamed per working step is smaller in accordance to the size of the hood.

Outdoors, a hood is positioned either manually or via tractor with a special pre-stressed 4 point suspension arm. Steaming time amounts to 30 min for a penetration down to 25 cm depth. Hereby a temperature of 90°C can be reached. In large stable glasshouses, the hoods are attached to tracks. They are lifted and moved by pneumatic cylinders. Small and medium-sized hoods up to 12m² are lifted manually using a tipping lever or moved electrically with special winches.

Combined Surface and Depth Injection of Steam (Sandwich Steaming)

Sandwich steaming machine model Sterilter constructed by Ferrari Costruzioni Meccaniche equipped with MSD/moeschle steam boiler

Sandwich steaming, which was developed in a project among DEIAFA, University of Turin (Italy, www.deiafa.unito.it) and Ferrari Costruzioni Meccaniche, represents a combination of depth and surface steaming, offers an efficient method to induce hot steam into the soil. The steam is simultaneously pushed into the soil from the surface and from the depth. For this purpose, the area, which must be equipped with a deep steaming injection system, is covered with a steaming hood. The steam enters the soil from the top and the bottom at the same time. Sheets are not suitable, since a high pressure up to 30 mm water column arises underneath the cover.

Sandwich steaming offers several advantages. On the one hand, application of energy can be increased to up to 120 kg steam per m²/h. In comparison to other steaming methods up to 30% energy savings can be achieved and the usage of fuel (e.g. heating oil) accordingly decreases. The increased application of energy leads to a quick heating of the soil which reduces the loss of heat. On the other hand, only half of the regular steaming time is needed.

Comparison of sandwich steaming with other steam injection methods relating to steam output and energy demand(*):

Steaming method	max. steam output	energy demand (*)
Sheet steaming	6 kg/m²h	about 100 kg steam/m³
Depth steaming (Sheet + vacuum)	14 kg/m²h	about 120 kg steam/m³
Hood steaming (Alu)	30 kg/m²h	about 80 kg steam/m³
Hood steaming (Steel)	50 kg/m²h	about 75 kg steam/m³
Sandwich steaming	120 kg/m²h	about 60 kg steam/m³

(*) in soil max 30% moisture

Clearly, Sandwich steaming reaches the highest steam output at the lowest energy demand.

Partially Integrated Sandwich Steaming

The partial integrated Sandwich steaming is an advanced combined method for steaming merely the areas which shall be planted and purposely leaving out those areas which shall not be used. In order to avoid risk of re-infection of steamed areas with pest from unsteamed areas, beneficial organisms can directly be injected into the hygenized soil via a soil activator (e.g. special compost). The partial sandwich steaming unlocks further potential savings in the steaming process.

Container / Stack Steaming

Stack steaming is used when thermically treating compost and substrates such as turf. Depending on the amount, the material to be steamed is piled up to 70 cm height in steaming boxes or in small dump trailers. Steam is evenly injected via manifolds. For huge amounts, steaming containers and soil boxes are used which are equipped with suction systems to improve steaming results. Midget amounts can be steamed in special small steaming devices.

The amount of soil steamed should be tuned in a way that steaming time amounts to at most 1.5 h in order to avoid large quantities of condensed water in the bottom layers of the soil.

Steam Output kg/h:	100	250	300	400	550	800	1000	1350	2000
m³/h about:	1.0-1.5	2.5-3.0	3.0-3.5	4.0-5.0	5.5-7.0	8.0-10.0	10.0-13.0	14.0-18.0	20.0-25.0

In light substrates, such as turf, the performance per hour is significantly higher.

History

Ancient civilizations in India and Egypt used steam, generated through the targeted usage of incident solar radiation on watered top soil, to sanitize and revive their arable land.

Modern soil steam sterilization was first discovered in 1888 (by Frank in Germany) and was first commercially used in the United States (by Rudd) in 1893 (Baker 1962). Since then, a wide variety of steam machines have been built to disinfest both commercial greenhouse and nursery field soils (Grossman and Liebman 1995). In the 1950s, for example, steam sterilization technologies expanded from disinfestation of potting soil and greenhouse mixes to commercial production of steam rakes and tractor-drawn steam blades for fumigating small acres of cut flowers and other high-value field crops (Langedijk 1959). Today, even more effective steam technologies are being developed.

Application of Hot Steam

- In horticulture as well as nurseries for sterilization of substrates and top soil

- In agriculture for sterilization and treatment of food waste for pig fattening and heating of molasses

- In mushroom cultivation for pasteurization of growing rooms, sterilization of top soil and combined application as heating

- In wineries as combination boiler for sterilization and cleaning of storage tanks, tempering of mash and for warm water generation.

Soil Solarization

Soil solarization is an environmentally friendly method of using solar power for controlling pests such as soilborne plant pathogens including fungi, bacteria, nematodes, and insect and mite pests along with weed seed and seedlings in the soil by mulching the soil and covering it with tarp, usually with a transparent polyethylene cover, to trap solar energy. It may also describe methods of decontaminating soil using sunlight or solar power. This energy causes physical, chemical, and biological changes in the soil.

Soil Disinfestation

Soil solarization (referred to as solar heating of the soil in early publications) is a relatively new soil disinfestation method, first described in extensive scientific detail by Katan et al. in 1976, presenting the results of a series of studies performed under field conditions, initiated in 1973, for controlling soilborne pathogens and weeds, mostly as a pre-planting soil treatment. Soil is mulched and then covered with transparent polyethylene during the hot season, thereby heating it and killing the pests.

Soil Decontamination

A 2008 study used a solar cell to generate an electric field for electrokinetic (EK) remediation of cadmium-contaminated soil. The solar cell could drive the electromigration of cadmium in contaminated soil, and the removal efficiency that was achieved by the solar cell was comparable with that achieved by conventional power supply.

In Korea, various remediation methods of soil slurry and groundwater contaminated with benzene at a polluted gas station site were evaluated, including a solar-driven, photocatalyzed reactor system along with various advanced oxidation processes (AOP). The most synergistic remediation method incorporated a solar light process with TiO_2 slurry and H_2O_2 system, achieving 98% benzene degradation, a substantial increase in the removal of benzene.

History

Attempts were made to use solar energy for controlling disease agents in soil and in plant material already in the ancient civilization of India. In 1939, Groashevoy, who used the term "solar energy for sand disinfection," controlled Thielaviopsis basicola upon heating the sand by exposure to direct sunlight.

Soil solarization is the third approach for soil disinfestation; the two other main approaches, soil steaming and fumigation; were developed at the end of the 19th century. The idea of solarization was based on observations by extension workers and farmers in the hot Jordan Valley, who noticed the intensive heating of the polyethylene-mulched soil. The involvement of biological control mechanisms in pathogen control and the possible implications were indicated in the first publication, noticing the very long effect of the treatment. In 1977, American scientists from the University of California at Davis reported the control of Verticillium in a cotton field, based on studies started in 1976, thus denoting, for the first time, the possible wide applicability of this method.

The use of polyethylene for soil solarization differs in principle from its traditional agricultural use. With solarization, soil is mulched during the hottest months (rather than the coldest, as in conventional plasticulture which is aimed at protecting the crop) in order to increase the maximal temperatures in an attempt to achieve lethal heat levels.

In the first 10 years following the influential 1976 publication, soil solarization was investigated in at least 24 countries and has been now been applied in more than 50, mostly in the hot regions, although there were some important exceptions. Studies have demonstrated effectiveness of solarization with various crops, including vegetables, field crops, ornamentals and fruit trees, against many pathogens, weeds and a soil arthropod. Those pathogens and weeds which are not controlled by solarization were also detected. The biological, chemical and physical changes that take in solarized soil during and after the solarization have been investigated, as well as the interaction of solarization with other methods of control. Long-term effects including biological control and increased growth response were verified in various climatic regions and soils, demonstrating the general applicability of solarization. Computerized simulation models have been developed to guide researchers and growers whether the ambient conditions of their locality are suitable for solarization.

Studies of the improvement of solarization by integrating it with other methods or by solarizing in closed glasshouses, or studies concerning commercial application by developing mulching machines were also carried out.

The use of solarization in existing orchards (e.g. controlling Verticillium in pistachio plantations) is an important deviation from the standard preplanting method and was reported as early as 1979.

Stale Seed Bed

A false or stale seed bed is a useful weed control technique which involves creating a seedbed some weeks before seed is due to be sown. Preparation of such seedbed makes sure that any weed seeds that have been disturbed and brought to the soil surface during cultivation will thus have a chance to germinate, and can then be hoed off or eliminated with the use of a flame weeder before sowing of the actual crop is carried out.

The technique can be utilized in early spring, when the weather is still too cold for proper seed germination. Several passes are made with a rototiller or plow, then weed seeds are allowed to germinate as weather permits.

By tilling, the farmer increases the chance of weed seed germination by the same method as one would for favorable vegetable/crops: the fine soil allows weed seed to grow rapidly by allowing the seed to open and the roots to spread easier than in compacted soil. Deep tilling will also bring dormant seed to the surface for germination. Some species of plant are known for seeds that can lay deeply buried in the soil for years before favorable conditions allow germination.

Timing is important, however. Weed seeds must be tilled/howed or otherwise destroyed before they themselves can create new seeds. By destroying them early, the farmer eliminates most of that season's annual weeds. Turning the dead weeds back into the soil also increases soil nutrient content, although this difference is slight.

In many cases, several tillings are done, perhaps every two weeks beginning in very early spring. This allows more and more weed seeds to germinate only to be killed off later. This eliminates more weeds, but care must be used to not delay planting of a desirable crop later than the crop needs for a successful season's growth.

After several years, most, if not all, weeds can be eliminated from the seed bank in the soil. In some cases the effect can be noticed in the same year the process is first carried out.

If the weed patch is vigorous enough, the farmer may mow the patch first before tilling. This allows for easier/quicker decomposition in the soil when the plants are turned under.

Some farmers are noted for applying a light and inexpensive fertilizer mix to the soil to cause even more weed seeds to germinate. Although the merits of this may seem a trifle at first, it can pay dividends in the long run by eliminating seeds earlier that otherwise would have sprouted in later years. This too is open to debate.

Aerial Application

Aerial application, or what was formerly referred to as crop dusting, involves spraying crops with crop protection products from an agricultural aircraft. Planting certain types of seed are also included in aerial application. The specific spreading of fertilizer is also known as *aerial topdressing* in some countries.

Agricultural aircraft are highly specialized, purpose-built aircraft. Today's agricultural aircraft are often powered by turbine engines of up to 1500 hp and can carry as much as 800 gallons of crop protection product. Helicopters are sometimes used, and some aircraft serve double duty as water bombers in areas prone to wildfires.(These aircraft are referred to as S.E.A.T. "single engine air tankers").

History

Aerial Seed Sowing 1906

The first known aerial application of agricultural materials was by John Chaytor, who in 1906 spread seed over a swamped valley floor in Wairoa, New Zealand, using a hot air balloon with mobile tethers. Aerial sowing of seed still continues to this day with cover crop applications and rice planting.

Crop Dusting 1921

Lt. Macready (right) and McCook Field engineer E. Dormoy (left) in front of the 1st crop duster airplane (August 3, 1921)

The first known use of a heavier-than-air machine to disperse products occurred on 3 August 1921. Crop dusting was developed under the joint efforts of the U.S. Agriculture Department, and the U.S. Army Signal Corps's research station at McCook Field in Dayton, Ohio. Under the direction of McCook engineer Etienne Dormoy, a United States Army Air Service Curtiss JN4 Jenny piloted by John A. Macready was modified at McCook Field to spread lead arsenate to kill catalpa sphinx caterpillars at a Catalapa farm near Troy, Ohio in the United States. The first test was considered highly successful. The first commercial operations were begun in 1924, by Huff-Daland Crop Dusting, which was co-founded by McCook Field test pilot Lt. Harold R. Harris. Use of insecticide and fungicide for crop dusting slowly spread in the Americas and to a lesser extent other nations in the 1930s. The name 'crop dusting' originated here, as actual dust was spread across the crops. Today, aerial applicators use liquid crop protection products in very small doses.

Top Dressing 1939–1946

Aerial topdressing, the spread of fertilizers such as superphosphate, was developed in New Zealand in the 1940s by members of the Ministry of Public Works and RNZAF, led by Alan Pritchard and Doug Campbell - unofficial experiments by individuals within the government led to funded research. Initially fertilizer and seed were dropped together (1939), using a window mounted chute on a Miles Whitney Straight, but by the end of the 1940s different mixtures of fertilizer were being distributed from hoppers installed in war surplus Grumman Avengers and C-47 Skytrains, as well as some privately operated de Havilland Tiger Moths in New Zealand, and the practise was being adopted experimentally in Australia and the United Kingdom.

Purpose Built Aircraft

In 1951, Leland Snow begins designing the first aircraft specifically built for aerial application, the S-1.

In 1957, The Grumman G-164 Ag-Cat is the first aircraft designed by a major company for ag aviation.

Water Bombing 1952

Aerial firefighting, or water bombing, was tested experimentally by Art Seller's Skyways air services in Canada in 1952 (dropping a mix of water, fertilizer and seed), and established in California in the mid-1950s.

Night Aerial Application 1973–Present

Aerial application at night is mostly liquid spray and is conducted in the Southwest U.S. deserts. The increased heat, scheduling conflicts with farm workers in the fields and honeybee activity reduced the effectiveness of spraying in daytime. In high temperature areas, the insects would travel down in plants in daytime and return to the top at night. The aircraft — both fixed wing, autogyros and helicopters — were equipped with lights, usually three sets: Work lights were high power and aimed or adjustable from the cockpit; wire lights were angled down for taxiing and wire or obstruction illumination; and turn lights were only turned on in the direction of the turn to allow safe operation on moonless nights where angle of entry or exit needed to be illuminated. Some aircraft were equipped with an elongated metal wing called a spreader, with inbuilt channels to direct the flow of dust such as sulfur, used on melons as a pesticide and soil amendment. Very little pesticide dust was used day or night in comparison to spray, because of the difficulty in drift control. Workers on the ground, called "flaggers", would use flashlights aimed at the aircraft to mark the swaths on the ground; later, GPS units replaced the flaggers due to new laws restricting use of human flaggers with some pesticides. GPS systems also provide precise guidance for the applicator.

Agricultural chemicals have also kept pace with advancements in technology, and have been influential in the growth of the agricultural aviation industry. In the 1930s Aerial Applicators arrived in the northern states to war against insect and disease pests which threatened fruit and vegetable crops. After World War II, the industry expanded into the western states where the development of new chemicals made possible the control of weeds and insects in cereal grain crops. Some of these new chemicals proved very useful in controlling various insects that carried diseases dangerous to humans. Countries that previously had no control over malaria and river blindness were provided with chemicals which helped save hundreds of thousands of lives and reduced the suffering of millions. All during the 1950s, crop production continued to rise and disease declined as a result of chemical controls.

Aerial application accounts for just under 20% of all applied crop protection products on commercial farms. The industry also provides firefighting and public health application services According to a 2012 NAAA survey, the five most common aerially treated crops are: corn, wheat/barley, soybeans, pastures/rangelands and alfalfa, but aerial application is used on many more crops grown in the U.S.

Approximately 1,350 aerial application businesses are in the U.S. and 1,430 non-operator pilots. 94% of aerial application business owners (operators) are also pilots. Aerial application businesses are located in 44 states – all but Connecticut, Hawaii, Nevada, Rhode Island, Vermont and West Virginia.

Today's ag aircraft use sophisticated precision application equipment such as: GPS (global positioning systems), GIS (geographical information systems), Aircraft Integrated Meteorological Measurement System (AIMMS), real time meteorological systems, flow control valves for variable-rate applications, single-boom shutoff valves and smokers to identify wind speed and direction.

According to the U.S. Bureau of Labor Statistics, in 2005 U.S. cropduster pilots earned an average annual wage of $63,210.

Unmanned Aerial Application

A Yamaha R-MAX, a UAV commonly used for aerial application in Japan.

Beginning in the late 1990s, unmanned aerial vehicles are also being used for agricultural spraying. This phenomenon started in Japan and South Korea, where mountainous terrain and relatively small family-owned farms required lower-cost and higher precision spraying. As of 2014, the use of UAV crop dusters, such as the Yamaha R-MAX, is being expanded to the United States for use in spraying of vineyards.

Irrigation

Irrigation is the method in which water is supplied to plants at regular intervals for agriculture. It is used to assist in the growing of agricultural crops, maintenance of landscapes, and revegetation of disturbed soils in dry areas and during periods of inadequate rainfall. Additionally, irrigation also has a few other uses in crop production, which include protecting plants against frost, suppressing weed growth in grain fields and preventing soil consolidation. In contrast, agriculture that relies only on direct rainfall is referred to as rain-fed or dry land farming.

Irrigation systems are also used for dust suppression, disposal of sewage, and in mining. Irrigation is often studied together with drainage, which is the natural or artificial removal of surface and sub-surface water from a given area.

Irrigation has been a central feature of agriculture for over 5,000 years and is the product of many cultures. Historically, it was the basis for economies and societies across the globe, from Asia to the Southwestern United States.

History

Animal-powered irrigation, Upper Egypt, ca. 1846

An example of an irrigation system common on the Indian subcontinent. Artistic impression on the banks of Dal Lake, Kashmir, India

Inside a karez tunnel at Turpan, Sinkiang

Archaeological investigation has identified as evidence of irrigation where the natural rainfall was insufficient to support crops for rainfed agriculture.

Perennial irrigation was practiced in the Mesopotamian plain whereby crops were regularly watered throughout the growing season by coaxing water through a matrix of small channels formed in the field.

irrigation in Tamil Nadu (India)

Ancient Egyptians practiced *Basin irrigation* using the flooding of the Nile to inundate land plots which had been surrounded by dykes. The flood water was held until the fertile sediment had settled before the surplus was returned to the watercourse. There is evidence of the ancient Egyptian pharaoh Amenemhet III in the twelfth dynasty (about 1800 BCE) using the natural lake of the Faiyum Oasis as a reservoir to store surpluses of water for use during the dry seasons, the lake swelled annually from flooding of the Nile.

The Ancient Nubians developed a form of irrigation by using a waterwheel-like device called a *sakia*. Irrigation began in Nubia some time between the third and second millennium BCE. It largely depended upon the flood waters that would flow through the Nile River and other rivers in what is now the Sudan.

In sub-Saharan Africa irrigation reached the Niger River region cultures and civilizations by the first or second millennium BCE and was based on wet season flooding and water harvesting.*Terrace irrigation* is evidenced in pre-Columbian America, early Syria, India, and China. In the Zana Valley of the Andes Mountains in Peru, archaeologists found remains of three irrigation canals radiocarbon dated from the 4th millennium BCE, the 3rd millennium BCE and the 9th century CE. These canals are the earliest record of irrigation in the New World. Traces of a canal possibly dating from the 5th millennium BCE were found under the 4th millennium canal. Sophisticated irrigation and storage systems were developed by the Indus Valley Civilization in present-day Pakistan and North India, including the reservoirs at Girnar in 3000 BCE and an early canal irrigation system from circa 2600 BCE. Large scale agriculture was practiced and an extensive network of canals was used for the purpose of irrigation.

Ancient Persia (modern day Iran) as far back as the 6th millennium BCE, where barley was grown in areas where the natural rainfall was insufficient to support such a crop. The Qanats, developed in ancient Persia in about 800 BCE, are among the oldest known irrigation methods still in use today. They are now found in Asia, the Middle East and North Africa. The system comprises a network of vertical wells and gently sloping tunnels driven into the sides of cliffs and steep hills to tap groundwater. The noria, a water wheel with clay pots around the rim powered by the flow of the stream (or by animals where the water source was still), was first brought into use at about this time, by Roman settlers in North Africa. By 150 BCE the pots were fitted with valves to allow smoother filling as they were forced into the water.

The irrigation works of ancient Sri Lanka, the earliest dating from about 300 BCE, in the reign of King Pandukabhaya and under continuous development for the next thousand years, were one of the most complex irrigation systems of the ancient world. In addition to underground canals, the Sinhalese were the first to build completely artificial reservoirs to store water. Due to their engineering superiority in this sector, they were often called 'masters of irrigation'. Most of these irrigation systems still exist undamaged up to now, in Anuradhapura and Polonnaruwa, because of the advanced and precise engineering. The system was extensively restored and further extended during the reign of King Parakrama Bahu (1153–1186 CE).

China

The oldest known hydraulic engineers of China were Sunshu Ao (6th century BCE) of the Spring and Autumn Period and Ximen Bao (5th century BCE) of the Warring States period, both of whom worked on large irrigation projects. In the Sichuan region belonging to the State of Qin of ancient China, the Dujiangyan Irrigation System was built in 256 BCE to irrigate an enormous area of farmland that today still supplies water. By the 2nd century AD, during the Han Dynasty, the Chinese also used chain pumps that lifted water from lower elevation to higher elevation. These were powered by manual foot pedal, hydraulic waterwheels, or rotating mechanical wheels pulled by oxen. The water was used for public works of providing water for urban residential quarters and palace gardens, but mostly for irrigation of farmland canals and channels in the fields.

Korea

In 15th century Korea, the world's first rain gauge, *uryanggye* (Korean:□□□), was invented in 1441. The inventor was Jang Yeong-sil, a Korean engineer of the Joseon Dynasty, under the active direction of the king, Sejong the Great. It was installed in irrigation tanks as part of a nationwide system to measure and collect rainfall for agricultural applications. With this instrument, planners and farmers could make better use of the information gathered in the survey.

North America

In North America, the Hohokam were the only culture to rely on irrigation canals to water their crops, and their irrigation systems supported the largest population in the Southwest by AD 1300. The Hohokam constructed an assortment of simple canals combined with weirs in their various agricultural pursuits. Between the 7th and 14th centuries, they also built and maintained extensive irrigation networks along the lower Salt and middle Gila rivers that rivaled the complexity of those used in the ancient Near East, Egypt, and China. These were constructed using relatively simple excavation tools, without the benefit of advanced engineering technologies, and achieved drops of a few feet per mile, balancing erosion and siltation. The Hohokam cultivated varieties of cotton, tobacco, maize, beans and squash, as well as harvested an assortment of wild plants. Late in the Hohokam Chronological Sequence, they also used extensive dry-farming systems, primarily to grow agave for food and fiber. Their reliance on agricultural strategies based on canal irrigation, vital in their less than hospitable desert environment and arid climate, provided the basis for the aggregation of rural populations into stable urban centers.

Present Extent

In the mid-20th century, the advent of diesel and electric motors led to systems that could pump groundwater out of major aquifers faster than drainage basins could refill them. This can lead to permanent loss of aquifer capacity, decreased water quality, ground subsidence, and other problems. The future of food production in such areas as the North China Plain, the Punjab, and the Great Plains of the US is threatened by this phenomenon.

Irrigation ditch in Montour County, Pennsylvania, off Strawberry Ridge Road

At the global scale, 2,788,000 km² (689 million acres) of fertile land was equipped with irrigation infrastructure around the year 2000. About 68% of the area equipped for irrigation is located in Asia, 17% in the Americas, 9% in Europe, 5% in Africa and 1% in Oceania. The largest contiguous areas of high irrigation density are found:

- In Northern India and Pakistan along the Ganges and Indus rivers

- In the Hai He, Huang He and Yangtze basins in China

- Along the Nile river in Egypt and Sudan

- In the Mississippi-Missouri river basin and in parts of California

Smaller irrigation areas are spread across almost all populated parts of the world.

Only eight years later in 2008, the scale of irrigated land increased to an estimated total of 3,245,566 km² (802 million acres), which is nearly the size of India.

Types of Irrigation

Basin flood irrigation of wheat

Irrigation of land in Punjab, Pakistan

Various types of irrigation techniques differ in how the water obtained from the source is distributed within the field. In general, the goal is to supply the entire field uniformly with water, so that each plant has the amount of water it needs, neither too much nor too little.

Surface Irrigation

In *surface* (*furrow*, *flood*, or *level basin*) irrigation systems, water moves across the surface of agricultural lands, in order to wet it and infiltrate into the soil. Surface irrigation can be subdivided into furrow, *borderstrip or basin irrigation*. It is often called *flood irrigation* when the irrigation results in flooding or near flooding of the cultivated land. Historically, this has been the most common method of irrigating agricultural land and still is in most parts of the world.

Where water levels from the irrigation source permit, the levels are controlled by dikes, usually plugged by soil. This is often seen in terraced rice fields (rice paddies), where the method is used to flood or control the level of water in each distinct field. In some cases, the water is pumped, or lifted by human or animal power to the level of the land. The field water efficiency of surface irrigation is typically lower than other forms of irrigation but has the potential for efficiencies in the range of 70% - 90% under appropriate management.

Localized Irrigation

Impact sprinkler head

Localized irrigation is a system where water is distributed under low pressure through a piped network, in a pre-determined pattern, and applied as a small discharge to each plant or adjacent to it. Drip irrigation, spray or micro-sprinkler irrigation and bubbler irrigation belong to this category of irrigation methods.

Subsurface Textile Irrigation

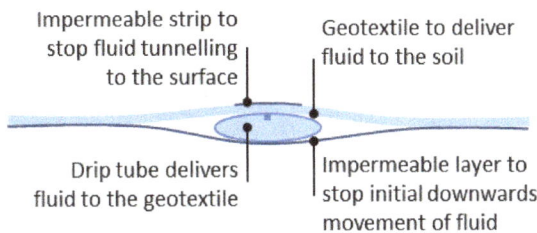

Impermeable strip to stop fluid tunnelling to the surface

Geotextile to deliver fluid to the soil

Drip tube delivers fluid to the geotextile

Impermeable layer to stop initial downwards movement of fluid

Diagram showing the structure of an example SSTI installation

Subsurface Textile Irrigation (SSTI) is a technology designed specifically for subsurface irrigation in all soil textures from desert sands to heavy clays. A typical subsurface textile irrigation system has an impermeable base layer (usually polyethylene or polypropylene), a drip line running along that base, a layer of geotextile on top of the drip line and, finally, a narrow impermeable layer on top of the geotextile. Unlike standard drip irrigation, the spacing of emitters in the drip pipe is not critical as the geotextile moves the water along the fabric up to 2 m from the dripper.

Drip Irrigation

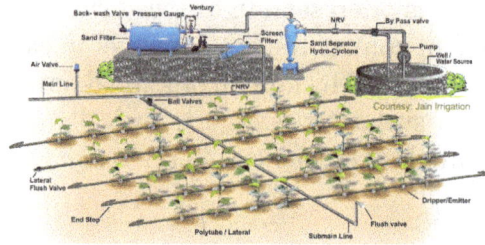

Drip irrigation layout and its parts

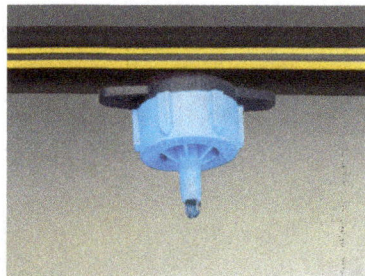

Drip irrigation – a dripper in action

Grapes in Petrolina, only made possible in this semi arid area by drip irrigation

Drip (or micro) irrigation, also known as trickle irrigation, functions as its name suggests. In this system water falls drop by drop just at the position of roots. Water is delivered at or near the root zone of plants, drop by drop. This method can be the most water-efficient method of irrigation, if managed properly, since evaporation and runoff are minimized. The field water efficiency of drip irrigation is typically in the range of 80 to 90 percent when managed correctly.

In modern agriculture, drip irrigation is often combined with plastic mulch, further reducing evaporation, and is also the means of delivery of fertilizer. The process is known as *fertigation*.

Deep percolation, where water moves below the root zone, can occur if a drip system is operated for too long or if the delivery rate is too high. Drip irrigation methods range from very high-tech and computerized to low-tech and labor-intensive. Lower water pressures are usually needed than for most other types of systems, with the exception of low energy center pivot systems and surface irrigation systems, and the system can be designed for uniformity throughout a field or for precise water delivery to individual plants in a landscape containing a mix of plant species. Although it is difficult to regulate pressure on steep slopes, pressure compensating emitters are available, so the field does not have to be level. High-tech solutions involve precisely calibrated emitters located along lines of tubing that extend from a computerized set of valves.

Irrigation Using Sprinkler Systems

Sprinkler irrigation of blueberries in Plainville, New York, United States

A traveling sprinkler at Millets Farm Centre, Oxfordshire, United Kingdom

In *sprinkler* or overhead irrigation, water is piped to one or more central locations within the field and distributed by overhead high-pressure sprinklers or guns. A system utilizing sprinklers, sprays, or guns mounted overhead on permanently installed risers is often referred to as a *solid-set* irrigation system. Higher pressure sprinklers that rotate are called *rotors* an are driven by a ball drive, gear drive, or impact mechanism. Rotors can be designed to rotate in a full or partial circle. Guns are similar to rotors, except that they generally operate at very high pressures of 40 to 130 lbf/in² (275 to 900 kPa) and flows of 50 to 1200 US gal/min (3 to 76 L/s), usually with nozzle diameters in the range of 0.5 to 1.9 inches (10 to 50 mm). Guns are used not only for irrigation, but also for industrial applications such as dust suppression and logging.

Sprinklers can also be mounted on moving platforms connected to the water source by a hose. Automatically moving wheeled systems known as *traveling sprinklers* may irrigate areas such as small farms, sports fields, parks, pastures, and cemeteries unattended. Most of these utilize a length of polyethylene tubing wound on a steel drum. As the tubing is wound on the drum powered by the irrigation water or a small gas engine, the sprinkler is pulled across the field. When the sprinkler arrives back at the reel the system shuts off. This type of system is known to most people as a "waterreel" traveling irrigation sprinkler and they are used extensively for dust suppression, irrigation, and land application of waste water.

Other travelers use a flat rubber hose that is dragged along behind while the sprinkler platform is pulled by a cable. These cable-type travelers are definitely old technology and their use is limited in today's modern irrigation projects.

Irrigation using Center Pivot

A small center pivot system from beginning to end

Center pivot irrigation

Center pivot irrigation is a form of sprinkler irrigation consisting of several segments of pipe (usually galvanized steel or aluminium) joined together and supported by trusses, mounted on wheeled towers with sprinklers positioned along its length. The system moves in a circular pattern and is fed with water from the pivot point at the center of the arc. These systems are found and used in all parts of the world and allow irrigation of all types of terrain. Newer systems have drop sprinkler heads as shown in the image that follows.

Most center pivot systems now have drops hanging from a u-shaped pipe attached at the top of the pipe with sprinkler head that are positioned a few feet (at most) above the crop, thus limiting evaporative losses. Drops can also be used with drag hoses or bubblers that deposit the water directly on the ground between crops. Crops are often planted in a circle to conform to the center pivot. This type of system is known as LEPA (Low Energy Precision Application). Originally, most center pivots were water powered. These were replaced by hydraulic systems (*T-L Irrigation*) and electric motor driven systems (Reinke, Valley, Zimmatic). Many modern pivots feature GPS devices.

Irrigation by Lateral Move (Side Roll, Wheel Line, Wheelmove)

A *series of pipes, each with a wheel* of about 1.5 m diameter permanently affixed to its midpoint, and sprinklers along its length, are coupled together. Water is supplied at one end using a large hose. After sufficient irrigation has been applied to one strip of the field, the hose is removed, the water drained from the system, and the assembly rolled either by hand or with a purpose-built mechanism, so that the sprinklers are moved to a different position across the field. The hose is reconnected. The process is repeated in a pattern until the whole field has been irrigated.

This system is less expensive to install than a center pivot, but much more labor-intensive to operate - it does not travel automatically across the field: it applies water in a stationary strip, must be drained, and then rolled to a new strip. Most systems use 4 or 5-inch (130 mm) diameter aluminum pipe. The pipe doubles both as water transport and as an axle for rotating all the wheels. A drive system (often found near the centre of the wheel line) rotates the clamped-together pipe sections as a single axle, rolling the whole wheel line. Manual adjustment of individual wheel positions may be necessary if the system becomes misaligned.

Wheel line systems are limited in the amount of water they can carry, and limited in the height of crops that can be irrigated. One useful feature of a lateral move system is that it consists of sections that can be easily disconnected, adapting to field shape as the line is moved. They are most often used for small, rectilinear, or oddly-shaped fields, hilly or mountainous regions, or in regions where labor is inexpensive.

Sub-Irrigation

Subirrigation has been used for many years in field crops in areas with high water tables. It is a method of artificially raising the water table to allow the soil to be moistened from below the plants' root zone. Often those systems are located on permanent grasslands in lowlands or river valleys and combined with drainage infrastructure. A system of pumping stations, canals, weirs and gates allows it to increase or decrease the water level in a network of ditches and thereby control the water table.

Sub-irrigation is also used in commercial greenhouse production, usually for potted plants. Water is delivered from below, absorbed upwards, and the excess collected for recycling. Typically, a solution of water and nutrients floods a container or flows through a trough for a short period of time, 10–20 minutes, and is then pumped back into a holding tank for reuse. Sub-irrigation in greenhouses requires fairly sophisticated, expensive equipment and management. Advantages are water and nutrient conservation, and labor-saving through lowered system maintenance and automation. It is similar in principle and action to subsurface basin irrigation.

Irrigation Automatically, Non-Electric using Buckets and Ropes

Besides the common manual watering by bucket, an automated, natural version of this also exists. Using plain polyester ropes combined with a prepared ground mixture can be used to water plants from a vessel filled with water.

The ground mixture would need to be made depending on the plant itself, yet would mostly consist of black potting soil, vermiculite and perlite. This system would (with certain crops) allow to save expenses as it does not consume any electricity and only little water (unlike sprinklers, water timers, etc.). However, it may only be used with certain crops (probably mostly larger crops that do not need a humid environment; perhaps e.g. paprikas).

Irrigation using Water Condensed from Humid Air

In countries where at night, humid air sweeps the countryside.Water can be obtained from the humid air by condensation onto cold surfaces. This is for example practiced in the vineyards at Lanzarote using stones to condense water or with various fog collectors based on canvas or foil sheets.

In-ground Irrigation

Most commercial and residential irrigation systems are "in ground" systems, which means that everything is buried in the ground. With the pipes, sprinklers, emitters (drippers), and irrigation valves being hidden, it makes for a cleaner, more presentable landscape without garden hoses or other items having to be moved around manually. This does, however, create some drawbacks in the maintenance of a completely buried system.

Most irrigation systems are divided into zones. A zone is a single irrigation valve and one or a group of drippers or sprinklers that are connected by pipes or tubes. Irrigation systems are divided into zones because there is usually not enough pressure and available flow to run sprinklers for an entire yard or sports field at once. Each zone has a solenoid valve on it that is controlled via wire by an irrigation controller. The irrigation controller is either a mechanical (now the "dinosaur" type) or electrical device that signals a zone to turn on at a specific time and keeps it on for a specified amount of time. "Smart Controller" is a recent term for a controller that is capable of adjusting the watering time by itself in response to current environmental conditions. The smart controller determines current conditions by means of historic weather data for the local area, a soil moisture sensor (water potential or water content), rain sensor, or in more sophisticated systems satellite feed weather station, or a combination of these.

When a zone comes on, the water flows through the lateral lines and ultimately ends up at the irrigation emitter (drip) or sprinkler heads. Many sprinklers have pipe thread inlets on the bottom of them which allows a fitting and the pipe to be attached to them. The sprinklers are usually installed with the top of the head flush with the ground surface. When the water is pressurized, the head will pop up out of the ground and water the desired area until the valve closes and shuts off that zone. Once there is no more water pressure in the lateral line, the sprinkler head will retract back into the ground. Emitters are generally laid on the soil surface or buried a few inches to reduce evaporation losses.

Water Sources

Irrigation is underway by pump-enabled extraction directly from the Gumti, seen in the background, in Comilla, Bangladesh.

Irrigation water can come from groundwater (extracted from springs or by using wells), from surface water (withdrawn from rivers, lakes or reservoirs) or from non-conventional sources like treated wastewater, desalinated water or drainage water. A special form of irrigation using surface water is spate irrigation, also called floodwater harvesting. In case of a flood (spate), water is diverted to normally dry river beds (wadis) using a network of dams, gates and channels and spread over large areas. The moisture stored in the soil will be used thereafter to grow crops. Spate irrigation areas are in particular located in semi-arid or arid, mountainous regions. While floodwater harvesting belongs to the accepted irrigation methods, rainwater harvesting is usually not considered as a form of irrigation. Rainwater harvesting is the collection of runoff water from roofs or unused land and the concentration of this.

Around 90% of wastewater produced globally remains untreated, causing widespread water pollution, especially in low-income countries. Increasingly, agriculture uses untreated wastewater as a source of irrigation water. Cities provide lucrative markets for fresh produce, so are attractive to farmers. However, because agriculture has to compete for increasingly scarce water resources with industry and municipal users, there is often no alternative for farmers but to use water polluted with urban waste, includ-ing sewage, directly to water their crops. Significant health hazards can result from using water loaded with pathogens in this way, especially if people eat raw vegetables that have been irrigated with the polluted water. The International Water Management Institute has worked in India, Pakistan, Vietnam, Ghana, Ethiopia, Mexico and other countries on various projects aimed at assessing and reducing risks of wastewater irrigation. They advocate a 'multiple-barrier' approach to wastewater use, where farmers are encouraged to adopt various risk-reducing behaviours. These include ceasing irrigation a few days before harvesting to allow pathogens to die off in the sunlight, applying water carefully so it does not contaminate leaves likely to be eaten raw, cleaning vegetables with disinfectant or allowing fecal sludge used in farming to dry before being used as a human manure. The World Health Organization has developed guidelines for safe water use.

There are numerous benefits of using recycled water for irrigation, including the low cost (when compared to other sources, particularly in an urban area), consistency of supply (regardless of season, climatic conditions and associated water restrictions), and general consistency of quality. Irrigation of recycled wastewater is also considered as a means for plant fertilization and particularly nutrient supplementation. This approach carries with it a risk of soil and water pollution through excessive wastewater application. Hence, a detailed understanding of soil water conditions is essential for effective utilization of wastewater for irrigation.

Efficiency

Modern irrigation methods are efficient enough to supply the entire field uniformly with water, so that each plant has the amount of water it needs, neither too much nor too little. Water use efficiency in the field can be determined as follows:

Young engineers restoring and developing the old Mughal irrigation system during the reign of the Mughal Emperor Bahadur Shah II

- Field Water Efficiency (%) = (Water Transpired by Crop ÷ Water Applied to Field) x 100

Until 1960s, the common perception was that water was an infinite resource. At that time, there were fewer than half the current number of people on the planet. People were not as wealthy as today, consumed fewer calories and ate less meat, so less water was needed to produce their food. They required a third of the volume of water we presently take from rivers. Today, the competition for water resources is much more intense. This is because there are now more than seven billion people on the planet, their consumption of water-thirsty meat and vegetables is rising, and there is increasing competition for water from industry, urbanisation and biofuel crops. To avoid a global water crisis, farmers will have to strive to increase productivity to meet growing demands for food, while industry and cities find ways to use water more efficiently.

Successful agriculture is dependent upon farmers having sufficient access to water. However, water scarcity is already a critical constraint to farming in many parts of the world. With regards to agriculture, the World Bank targets food production and water management as an increasingly global issue that is fostering a growing debate. Physical water scarcity is where there is not enough water to meet all demands, including that needed for ecosystems to function effectively. Arid regions frequently suffer from physical water scarcity. It also occurs where water seems abundant but where resources are over-committed. This can happen where there is overdevelopment of hydraulic infrastructure, usually for irrigation. Symptoms of physical water scarcity include environmental degradation and declining groundwater. Economic scarcity, meanwhile, is caused by a lack of investment in water or insufficient human capacity to satisfy the demand for water. Symptoms of economic water scarcity include a lack of infrastructure, with people often having to fetch water from rivers for domestic and agricultural uses. Some 2.8 billion people currently live in water-scarce areas.

Technical Challenges

Irrigation schemes involve solving numerous engineering and economic problems while minimizing negative environmental impact.

- Competition for surface water rights.

- Overdrafting (depletion) of underground aquifers.

- Ground subsidence (e.g. New Orleans, Louisiana)

- Underirrigation or irrigation giving only just enough water for the plant (e.g. in drip line irrigation) gives poor soil salinity control which leads to increased soil salinity with consequent buildup of toxic salts on soil surface in areas with high evaporation. This requires either leaching to remove these salts and a method of drainage to carry the salts away. When using drip lines, the leaching is best done regularly at certain intervals (with only a slight excess of water), so that the salt is flushed back under the plant's roots.

- Overirrigation because of poor distribution uniformity or management wastes water, chemicals, and may lead to water pollution.

- Deep drainage (from over-irrigation) may result in rising water tables which in some instances will lead to problems of irrigation salinity requiring watertable control by some form of subsurface land drainage.

- Irrigation with saline or high-sodium water may damage soil structure owing to the formation of alkaline soil

- Clogging of filters: It is mostly algae that clog filters, drip installations and nozzles. UV and ultrasonic method can be used for algae control in irrigation systems.

References

- Crafts, Alden S. (1975). Modern Weed Control. Berkeley, California: University of California Press. pp. 110–117. ISBN 0-520-02733-7.

- Rao, V.S. (2000). Principles of Weed Science. Enfield, New Hampshire: Science Publishers, Inc. pp. 39–48. ISBN 1-57808-069-X.

- Turgeon; et al. (2009). Weed Control in Turf and Ornamentals. Upper Saddle River, New Jersey: Pearson Education, Inc. pp. 127–129. ISBN 0-13-159122-3.

- Hynes, Erin (1995). Controlling Weeds. Emmaus, Pennsylvania: Rodale Press. pp. 26–30. ISBN 0-87596-667-5.

- G. Mokhtar (1981-01-01). Ancient civilizations of Africa. Books.google.com. Unesco. International Scientific Committee for the Drafting of a General History of Africa. p. 309. ISBN 9780435948054. Retrieved 2012-06-19.

- Richard Bulliet, Pamela Kyle Crossley, Daniel Headrick, Steven Hirsch. Pages 53-56 (2008-06-18). The Earth and Its Peoples, Volume I: A Global History, to 1550. Books.google.com. ISBN 0618992383. Retrieved 2012-06-19.

- Frenken, K. (2005). Irrigation in Africa in figures – AQUASTAT Survey – 2005 (PDF). Food and Agriculture Organization of the United Nations. ISBN 92-5-105414-2. Retrieved 2007-03-14.

- Drainage Manual: A Guide to Integrating Plant, Soil, and Water Relationships for Drainage of Irrigated Lands. Interior Dept., Bureau of Reclamation. 1993. ISBN 0-16-061623-9.

- Mader, Shelli (May 25, 2010). "Center pivot irrigation revolutionizes agriculture". The Fence Post Magazine. Retrieved June 6, 2012.

- "Polyester ropes natural irrigation technique". Entheogen.com. Archived from the original on April 12, 2012. Retrieved 2012-06-19.

- "DIY instructions for making self-watering system using ropes". Instructables.com. 2008-03-17. Retrieved 2012-06-19.

- Bell, Carl; Dean Lehman (June 2005). "Best Management Practices for Vegetation Management" (PDF). Los Angeles County Weed Management Area. Retrieved 17 Feb 2012.

- "Managing Invasive Plants: Concepts, Principles, and Practices". United States Fish & Wildlife Service. Retrieved 19 Feb 2012.

- "Africa, Emerging Civilizations In Sub-Sahara Africa. Various Authors; Edited By: R. A. Guisepi". History-world.org. Retrieved 2012-06-19.

- "Reengaging in Agricultural Water Management: Challenges and Options". The World Bank. pp. 4–5. Retrieved 2011-10-30.

- "Free articles and software on drainage of waterlogged land and soil salinity control in irrigated land". Retrieved 2010-07-28.

4

Pest Control: Processes and Techniques

This segment carefully elaborates the basic concepts of pest control to provide a complete understanding. Processes and techniques, like mechanical pest control, physical pest control, fumigation, crop rotation and integrated pest management are explained in a critical and systematic manner. This chapter is an overview of the subject matter incorporating all the major aspects of weed and pest control.

Mechanical Pest Control

Mechanical pest control is the management and control of pests using physical means such as fences, barriers or electronic wires. It includes also weeding and change of temperature to control pests. Many farmers at the moment are trying to find sustainable ways to remove pests without harming the ecosystem.

Methods

Handpicking

The use of human hands to remove harmful insects or other toxic material is often the most common action by gardeners. It is also classified as the most direct and the quickest way to remove clearly visible pests. However, it also has equal disadvantages as it must be performed before damage to the plant has been done and before the key development of insects.

Mechanical Traps

Mechanical traps or physical attractants are used in three main ways: to efficiently trap insects, to kill them or to estimate how much many insects there are in the total landmass using sampling method. However, some traps are expensive to produce and can end up benefiting insects rather than harming them.

Differences from Integrated Pest Control

Integrated pest control refers to the use of any means to control pests once they reach unacceptable levels. Mechanical pest control is but a minor part of integrated pest control. It means only the use of physical means to control pests

Physical Pest Control

Physical Pest Control is a method of getting rid of insects and small rodents by re-moving, attacking, or setting up barriers that will prevent further destruction of one's plants. These methods are used primarily for crop growing, but some methods can be applied to homes as well.

Methods

Barriers

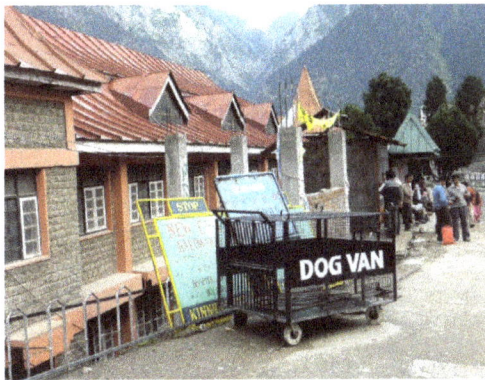

Dog control van, Rekong Peo, Himachal Pradesh, India

Row covers are useful for keeping insects out of one's plants, typically used for horti-cultural crops. They are made out of either plastic or polyester. They are made thin and light to allow plants to still absorb sunshine and water from the air.

Diatomaceous earth, made from fossilized and pulverized silica shells, can be used in order to damage the protective cuticle layer of insects that have them, such as ants. When this layer is damaged, the insects become vulnerable to drying out. Unfortunate-ly, the effectiveness of Diatomaceous earth decreases if it is wet. Therefore, it must be used often. This method was used back in the 1930s and 1940s when farmers would run dust over their fields. This would have the very same effect as diatomaceous earth.

Traps

Fly paper or sticky boards are devices used in order to capture insects as they land upon the surface of the trap. They are covered in a substance that attracts insects, but are ac-tually very sticky or poisonous. These traps are commonly used for flies or leafhoppers.

Trap strips are crops that are grown on fields with the intention of using them to attract insects and not have insects infest the other crops that are being grown. The insects can then be dealt with much more easily than if they were to have been spread throughout an entire field. Trap strips are very useful for dealing with the wheat stem sawfly. The sawflies will go only as far as they need to in order to plant their eggs. If solid stemmed plants are planted around the a crop field, then that's where the sawflies will go and the sawflies' larvae can't survive in the solid stem.

Fire

For farmers, fire has been a powerful technique used to destroy insect breeding grounds. It is used to burn the top of the soil in order to kill the insects that lie there. Unfortunately, this can present some drawbacks. Fire can make the soil much less effective or get rid of the insects that are beneficial to the plants. Also, there is no guarantee that it will actually solve the pest problems since there may be larvae below the surface of the soil.

Temperature Control

Placing produce inside of cold storage containers lengthens how long the produce lasts while also hindering the growth of insects inside of them. Another method to use is to heat, as it will kill the insect larvae in certain types of produce. An example would be with mangoes, where they are placed into a hot water bath in order to kill any eggs and larvae.

Large Scale Usage

On a much larger scale, physical control methods become much less effective because of the time that must be invested into it and because it is likely to be less economical. For example, taking care of a single tree is simple, but taking care of 500, like on a farm, would be impossible using physical control.

Fumigation

Fumigation is a method of pest control that completely fills an area with gaseous pesticides—or fumigants—to suffocate or poison the pests within. It is used to control pests in buildings (structural fumigation), soil, grain, and produce, and is also used during processing of goods to be imported or exported to prevent transfer of exotic organisms. This method also affects the structure itself, affecting pests that inhabit the physical structure, such as woodborers and drywood termites.

Process

Fumigation generally involves the following phases: First the area intended to be fumigated is usually covered to create a sealed environment; next the fumigant is released into the space to be fumigated; then, the space is held for a set period while the fumigant gas percolates through the space and acts on and kills any infestation in the product, next the space is ventilated so that the poisonous gases are allowed to escape from the space, and render it safe for humans to enter. If successful, the fumigated area is now safe and pest free.

Tent Fumigation (Tenting)

Structural fumigation techniques differ from building to building, but in houses a rubber tent is often placed over the entire house while the pesticides are being released into the vacant residence. This process is called tent fumigation or "tenting". The sealed tent concentrates the poisonous gases and prevents them from escaping into the neighborhood. The process can take up to a week depending on the fumigant used, which is in turn dependent on the severity of infestation and size of the building.

Chemicals

Methyl bromide was among the most widely used fumigants until its production and use was restricted by the Montreal Protocol due to its role in ozone depletion.

Widely used fumigants include:

- 1,3-dichloropropene
- dazomet
- chloropicrin
- DBCP, prior to 1985
- formaldehyde
- hydrogen cyanide
- iodoform
- methyl isocyanate
- phosphine
- sulfuryl fluoride

Safety

Fumigation is a hazardous operation. Generally it is a legal requirement that the operator who carries out the fumigation operation holds official certification to perform the fumigation as the chemicals used are toxic to most forms of life, including humans.

Post operation ventilation of the area is a critical safety aspect of fumigation. It is important to distinguish between the pack or source of the fumigant gas and the environment which has been fumigated. While the fumigant pack may be safe and spent, the space will still hold the fumigant gas until it has been ventilated.

Electronic Pest Control

Electronic pest control is the name given to the use of any of the several types of electrically powered devices designed to repel or eliminate pests, usually rodents or insects. Since these devices are not regulated under the Federal Insecticide, Fungicide and Rodenticide Act (FIFRA) in the United States, the US EPA does not require the same kind of efficacy testing that it does for chemical pesticides.

Types of Devices

There are two types of electronic pest control devices widely available: electromagnetic and ultrasonic.

Electromagnetic

Electromagnetic ("EM") pest repelling devices claim to affect the nervous system of ants, mice, spiders, and other rodents. There have been similar studies on effects of EM radiation emitted by cellphones on humans.

Ultrasonic

Ultrasonic devices operate through emitting short wavelength, high frequency sound waves that are too high in pitch to be heard by the human ear (all frequencies greater than 20,000 Hz). Humans are unable to hear sounds higher than 20 kHz due to physiological limitations of the cochlea. Some animals, such as bats, dogs, and rodents, can hear well into the ultrasonic range. Some insects, such as grasshoppers and locusts, can detect frequencies from 50,000 Hz to 100,000 Hz, and lacewings and moths can detect ultrasound as high as 240,000 Hz produced by insect-hunting bats. Contrary to popular belief, birds cannot hear ultrasonic sound. Some smartphone applications attempt to use this technology to produce high frequency sounds to repel mosquitoes and other insects, but the claims of effectiveness of these application and of ultrasonic control of mosquitoes in general has been questioned.

Insects detect sound by special hairs or sensilla located on the antennae (mosquitoes) or genitalia (cockroaches), or by more complicated tympanal organs (butterflies, grasshoppers, locusts, and moths).

Radio Wave Pest Control

The concept of radio wave (RW) or radio frequency (RF) to control the behavior of living organisms has shown promise. According to Drs. Juming Tang and Shaojin Wang at Washington State University (WSU) with colleagues at the University of California-Davis and USDA's Agricultural Research Service in Parlier, California, since RF energy generates heat through agitation of bound water molecules, it generates heat through ionic conduction and agitation of free water molecules in insects. As a result, more thermal energy is converted in insects.

RF treatments control insect pests without negatively affecting food stuffs and storage locations. RF treatments may serve as a non-chemical alternative to chemical fumigants for post-harvest pest control in commodities (such as almonds, pecans, pistachios, lentils, peas, and soybeans), reducing the long-term impact on the environment, human health, and competitiveness of agricultural industries.

Effects on Pests

Studies

"Ultrasound and Arthropod Pest Control" (2001), an extensive Kansas State University study, confirmed that ultrasonic sound devices do have both a repellent effect as well as reduces mating and reproduction of insects. However, the results were mixed, and ultrasonic sound had little or no effect on some pests. Ultrasonic devices were highly effective on crickets, while the same devices had little repellent effect on cockroaches. Additionally, the results were mixed: some devices were effective, while others had no effect depending on the test subject. The study also concluded there was no effect on ants or spiders in any of the tests. They concluded, based on the mixed results, that more research is needed to improve these devices.

A 2002 study sponsored by Genesis Laboratories, Inc. (the maker of the Pest-A-Cator/Riddex series of electronic repellent devices) does lend some credence to the ability of electronic repellent devices to repel certain pests in controlled environments. "Preliminary study of white-footed mice behavior in the test apparatus demonstrated a significant preference for the non-activated chamber among both sexes."

In 2003, the Federal Trade Commission required Global Instruments, the maker of the Pest-A-Cator/Riddex series of electromagnetic pest control devices, to discontinue any claims for their efficacy until they are backed by credible scientific evidence. This ban continues to be in effect.

In 2009, Victor Pest obtained positive results from independent researchers which resulted in two ultrasonic devices' being granted registration by the Canadian EPA (PMRA). The results from the tests were: the device "successfully repelled the rodents from the protected area in 13 of the 17 sites. This represents a 81.3% success rate...the average number of days before rodent activity was stopped was six days".

Effects

Effects on Cockroaches

Cockroaches respond to electronic pest control devices by moving about a bit more than usual, but don't appear eager to escape from the sound waves. This includes devices that emit a uniform frequency as well as those that emit changing frequencies of ultrasound. Researchers were able to use the increased cockroach activity to good effect by increasing the rate at which they caught the roaches in sticky traps.

Effects on Mosquitoes

A 2007 review article examined 10 field studies, in which ultrasonic repellent devices had been put to the test, and concluded they "have no effect on preventing mosquito bites" and "should not be recommended or used". It goes on: "Given these findings from 10 carefully conducted studies, it would not be worthwhile to conduct further research on EMRs [electronic mosquito repellents] in preventing mosquitoes biting or in trying to prevent the acquisition of malaria".

Bart Knols, an entomologist who chairs the advisory board of the Dutch Malaria Foundation and edits the website Malaria World, claims there is "no scientific evidence whatsoever" that ultrasound repels mosquitoes.

In 2005, the British consumer magazine *Holiday* reported the results of its test of a range of mosquito deterrents. The magazine's editor Lorna Cowan described the four appliances that used a buzzer as "a shocking waste of money" which "should be removed from sale". One, the Lovebug, a ladybird-shaped gadget designed to be clipped onto a baby's cot or child's pushchair - was singled out as a particular cause for concern, because of the likelihood that parents would trust it to keep mosquitoes away, and their children would be hurt as a result. (The Lovebug is still readily available in Europe, though it was withdrawn from the US market after the Federal Trade Commission reprimanded the manufacturer Prince Lionheart.)

Effects on Rodents

Rodents adjust to the ultrasound (or any new sound) and eventually ignore it. At best, ultrasonic waves have only a partial or temporary effect on rodents. Numerous studies have rejected ultrasonic sound as a practical means of rodent control. Tests of commercial ultrasonic devices have indicated that rodents may be repelled from the immediate area of the ultrasound device for a few minutes to a few days, but they will nearly always return and resume normal activities. Other tests have shown that the degree of repellance depends on the frequency, intensity, and pre-existing condition of the rodent infestation. The intensity of such sounds must be so great that damage to humans or domestic animals would also be likely; commercial ultrasonic pest control devices do not produce sounds of such intensity.

Safety

Professor Tim Leighton at the Institute of Sound and Vibration Research], University of Southampton, U.K. produced an 83-page paper entitled "What is Ultrasound?"

(2007), in which he expressed concern about the growth in commercial products which exploit the discomforting effects of in-air ultrasound (to pests for whom it is within their audible frequency range, or to humans for whom it is not, but who can experience unpleasant subjective effects and, potentially, shifts in the hearing threshold). Leighton claims that commercial products are often advertised with cited levels which cannot be critically accepted due to lack of accepted measurement standards for ultrasound in air, and little understanding of the mechanism by which they may represent a hazard.

The UK's independent Advisory Group on Non-ionising Radiation (AGNIR) produced a 180-page report on the health effects of human exposure to ultrasound and infrasound in 2010. The UK Health Protection Agency (HPA) published their report, which recommended an exposure limit for the general public to airborne ultrasound sound pressure levels (SPL) of 70 dB (at 20 kHz), and 100 dB (at 25 kHz and above).

Biological Pest Control

Biological control is a method of controlling pests such as insects, mites, weeds and plant diseases using other organisms. It relies on predation, parasitism, herbivory, or other natural mechanisms, but typically also involves an active human management role. It can be an important component of integrated pest management (IPM) programs.

There are three basic types of biological pest control strategies: importation (sometimes called classical biological control), in which a natural enemy of a pest is introduced in the hope of achieving control; augmentation, in which locally-occurring natural enemies are bred and released to improve control; and conservation, in which measures are taken to increase natural enemies, such as by planting nectar-producing crop plants in the borders of rice fields.

Natural enemies of insect pests, also known as biological control agents, include predators, parasitoids, and pathogens. Biological control agents of plant diseases are most often referred to as antagonists. Biological control agents of weeds include seed predators, herbivores and plant pathogens.

Biological control can have side-effects on biodiversity through predation, parasitism, pathogenicity, competition, or other attacks on non-target species, especially when a species is introduced without thorough understanding of the possible consequences.

Types of Biological Pest Control

There are three basic biological pest control strategies: importation (classical biological control), augmentation and conservation.

Importation

Importation or classical biological control involves the introduction of a pest's natural enemies to a new locale where they do not occur naturally. Early instances were often unofficial and not based on research, and some introduced species became serious pests themselves.

Rodolia cardinalis, the vedalia beetle, was imported to Australia in the 19th century, successfully controlling cottony cushion scale.

To be most effective at controlling a pest, a biological control agent requires a colonizing ability which allows it to keep pace with the spatial and temporal disruption of the habitat. Control is greatest if the agent has temporal persistence, so that it can maintain its population even in the temporary absence of the target species, and if it is an opportunistic forager, enabling it to rapidly exploit a pest population.

Joseph Needham noted a Chinese text dating from 304 AD, *Records of the Plants and Trees of the Southern Regions*, by Hsi Han, which describes mandarin oranges protected by large reddish-yellow citrus ants which attack and kill insect pests of the orange trees. The citrus ant (*Oecophylla smaragdina*) was rediscovered in the 20th century, and since 1958 has been used in China to protect orange groves.

One of the earliest successes in the west was in controlling *Icerya purchasi* (cottony cushion scale) in Australia, using a predatory insect *Rodolia cardinalis* (the vedalia beetle). This success was repeated in California using the beetle and a parasitoid fly, *Cryptochaetum iceryae*.

Prickly pear cacti were introduced into Queensland, Australia as ornamental plants. They quickly spread to cover over 25 million hectares of Australia. Two control agents were used to help control the spread of the plant, the cactus moth *Cactoblastis cactorum*, and *Dactylopius* scale insects.

Damage from *Hypera postica*, the alfalfa weevil, a serious introduced pest of forage, was substantially reduced by the introduction of natural enemies. 20 years after their introduction the population of weevils in the alfalfa area treated for alfalfa weevil in the Northeastern United States remained 75 percent down.

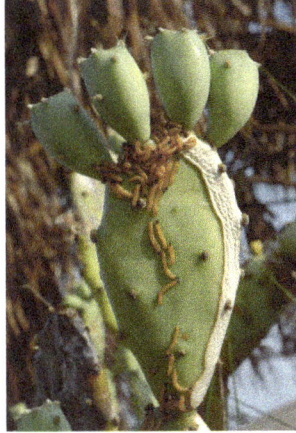

Cactoblastis cactorum larvae feeding on *Opuntia* cacti

The invasive species *Alternanthera philoxeroides* (alligator weed) was controlled in Florida (U.S.) by introducing alligator weed flea beetle.

Alligator weed was introduced to the United States from South America. It takes root in shallow water, interfering with navigation, irrigation, and flood control. The alligator weed flea beetle and two other biological controls were released in Florida, enabling the state to ban the use of herbicides to control alligator weed three years later. Another aquatic weed, the giant salvinia (*Salvinia molesta*) is a serious pest, covering waterways, reducing water flow and harming native species. Control with the salvinia weevil (*Cyrtobagous salviniae*) is effective in warm climates, and in Zimbabwe, a 99% control of the weed was obtained over a two-year period.

Small commercially reared parasitoidal wasps, *Trichogramma ostriniae*, provide limited and erratic control of the European corn borer (*Ostrinia nubilalis*), a serious pest. Careful formulations of the bacterium *Bacillus thuringiensis* are more effective.

The population of *Levuana iridescens*, the Levuana moth, a serious coconut pest in Fiji, was brought under control by a classical biological control program in the 1920s.

Augmentation

Augmentation involves the supplemental release of natural enemies, boosting the naturally occurring population. In inoculative release, small numbers of the control agents are released at intervals to allow them to reproduce, in the hope of setting up longer-term control, and thus keeping the pest down to a low level, constituting prevention rather than cure. In inundative release, in contrast, large numbers are released in the hope of rapidly reducing a damaging pest population, correcting a problem that has already arisen. Augmentation can be effective, but is not guaranteed to work, and relies on understanding of the situation.

Hippodamia convergens, the convergent lady beetle, is commonly sold for biological control of aphids.

An example of inoculative release occurs in greenhouse production of several crops. Periodic releases of the parasitoid, *Encarsia formosa*, are used to control greenhouse whitefly, while the predatory mite *Phytoseiulus persimilis* is used for control of the two-spotted spider mite.

The egg parasite *Trichogramma* is frequently released inundatively to control harmful moths. Similarly, *Bacillus thuringiensis* and other microbial insecticides are similarly used in large enough quantities for a rapid effect. Recommended release rates for *Trichogramma* in vegetable or field crops range from 5,000 to 200,000 per acre (1 to 50 per square metre) per week according to the level of pest infestation. Similarly, entomopathogenic nematodes are released at rates of millions and even billions per acre for control of certain soil-dwelling insect pests.

Conservation

The conservation of existing natural enemies in an environment is the third method of biological pest control. Natural enemies are already adapted to the habitat and to the target pest, and their conservation can be simple and cost-effective, as when nectar-producing crop plants are grown in the borders of rice fields. These provide nectar to support parasitoids and predators of planthopper pests and have been demonstrated to be so effective (reducing pest densities by 10- or even 100-fold) that farmers sprayed 70% less insecticides, enjoyed yields boosted by 5%, and this led to an economic advantage of 7.5%. Predators of aphids were similarly found to be present in tussock grasses by field boundary hedges in England, but they spread too slowly to reach the centres of fields. Control was improved by planting a metre-wide strip of tussock grasses in field centres, enabling aphid predators to overwinter there.

An inverted flowerpot filled with straw to attract earwigs

Cropping systems can be modified to favor natural enemies, a practice sometimes referred to as habitat manipulation. Providing a suitable habitat, such as a shelterbelt, hedgerow, or beetle bank where beneficial insects can live and reproduce, can help ensure the survival of populations of natural enemies. Things as simple as leaving a layer of fallen leaves or mulch in place provides a suitable food source for worms and provides a shelter for insects, in turn being a food source for such beneficial mammals as hedgehogs and shrews. Compost piles and stacks of wood can provide shelter for invertebrates and small mammals. Long grass and ponds support amphibians. Not removing dead annuals and non-hardy plants in the autumn allows insects to make use of their hollow stems during winter. In California, prune trees are sometimes planted in grape vineyards to provide an improved overwintering habitat or refuge for a key grape pest parasitoid. The providing of artificial shelters in the form of wooden caskets, boxes or flowerpots is also sometimes undertaken, particularly in gardens, to make a cropped area more attractive to natural enemies. For example, earwigs are natural predators which can be encouraged in gardens by hanging upside-down flowerpots filled with straw or wood wool. Green lacewings can be encouraged by using plastic bottles with an open bottom and a roll of cardboard inside. Birdhouses enable insectivorous birds to nest; the most useful birds can be attracted by choosing an opening just large enough for the desired species.

Biological Control Agents

Predators

Predators are mainly free-living species that directly consume a large number of prey during their whole lifetime. Ladybugs, and in particular their larvae which are active between May and July in the northern hemisphere, are voracious predators of aphids, and also consume mites, scale insects and small caterpillars. The spotted lady beetle (*Coleomegilla maculata*) is also able to feed on the eggs and larvae of the Colorado potato beetle (*Leptinotarsa decemlineata*).

Lacewings are available from biocontrol dealers.

The larvae of many hoverfly species principally feed upon greenfly (aphids), one larva devouring up to 400 in its lifetime. Their effectiveness in commercial crops has not been studied.

Predatory *Polistes* wasp looking for bollworms or other caterpillars on a cotton plant

Several species of entomopathogenic nematode are important predators of insect and other invertebrate pests. *Phasmarhabditis hermaphrodita* is a microscopic nematode that kills slugs. Its complex life cycle include a free-living, infective stage in the soil where it becomes associated with a pathogenic bacteria such as *Moraxella osloensis*. The nematode enters the slug through the posterior mantle region, thereafter feeding and reproducing inside, but it is the bacteria that kill the slug. The nematode is available commercially in Europe and is applied by watering onto moist soil.

Species used to control spider mites include the predatory mites *Phytoseiulus persimilis, Neoseilus californicus,* and *Amblyseius cucumeris*, the predatory midge *Feltiella acarisuga*, and a ladybird *Stethorus punctillum*. The bug *Orius insidiosus* has been successfully used against the two-spotted spider mite and the western flower thrips (*Frankliniella occidentalis*).

Parasitoids

Parasitoids lay their eggs on or in the body of an insect host, which is then used as a food for developing larvae. The host is ultimately killed. Most insect parasitoids are wasps or flies, and may have a very narrow host range. The most important groups are the ichneumonid wasps, which prey mainly on caterpillars of butterflies and moths; braconid wasps, which attack caterpillars and a wide range of other insects including greenfly; chalcid wasps, which parasitize eggs and larvae of greenfly, whitefly, cabbage caterpillars, and scale insects; and tachinid flies, which parasitize a wide range of insects including caterpillars, adult and larval beetles, and true bugs.

Encarsia formosa was one of the first biological control agents developed.

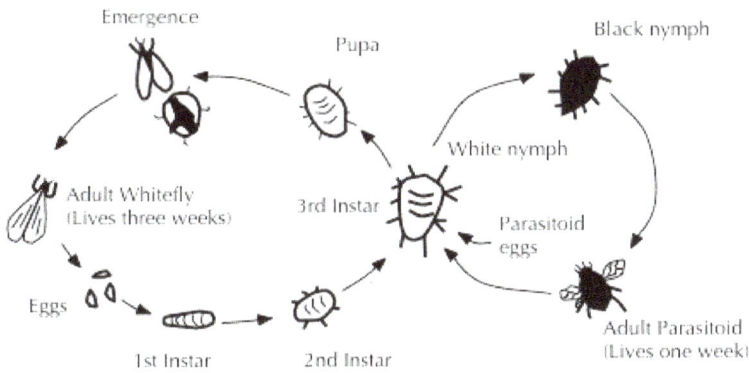

Life cycles of Greenhouse whitefly and its parasitoid wasp *Encarsia formosa*

Encarsia formosa is a small predatory chalcid wasp which is a parasitoid of whitefly, a sap-feeding insect which can cause wilting and black sooty moulds in glasshouse vegetable and ornamental crops. It is most effective when dealing with low level infestations, giving protection over a long period of time. The wasp lays its eggs in young whitefly 'scales', turning them black as the parasite larvae pupates. *Gonatocerus ashmeadi* (Hymenoptera: Mymaridae) has been introduced to control the glassy-winged sharpshooter *Homalodisca vitripennis* (Hemipterae: Cicadellidae) in French Polynesia and has successfully controlled ~95% of the pest density.

Parasitoids are among the most widely used biological control agents. Commercially, there are two types of rearing systems: short-term daily output with high production of parasitoids per day, and long-term low daily output with a range in production of 4-1000million female parasitoids per week. Larger production facilities produce on a yearlong basis, whereas some facilities produce only seasonally. Rearing facilities are usually a significant distance from where the agents are to be used in the field, and transporting the parasitoids from the point of production to the point of use can pose problems. Shipping conditions can be too hot, and even vibrations from planes or trucks can adversely affect parasitoids.

Pathogens

Pathogenic micro-organisms include bacteria, fungi, and viruses. They kill or debilitate their host and are relatively host-specific. Various microbial insect diseases occur naturally, but may also be used as biological pesticides. When naturally occurring, these outbreaks are density-dependent in that they generally only occur as insect populations become denser.

Bacteria

Bacteria used for biological control infect insects via their digestive tracts, so they offer only limited options for controlling insects with sucking mouth parts such as aphids and scale insects. *Bacillus thuringiensis* is the most widely applied species of bacteria used for biological control, with at least four sub-species used against Lepidopteran (moth, butterfly), Coleopteran (beetle) and Dipteran (true fly) insect pests. The bacterium is available in sachets of dried spores which are mixed with water and sprayed onto vulnerable plants such as brassicas and fruit trees. *B. thuringiensis* has also been incorporated into crops, making them resistant to these pests and thus reducing the use of pesticides. The bacterium *Paenibacillus popilliae* causes milky spore disease has been found useful in the control of Japanese beetle, killing the larvae. It is very specific to its host species and is harmless to vertebrates and other invertebrates.

Fungi

Green peach aphid, a pest in its own right and a vector of plant viruses, killed by the fungus *Pandora neoaphidis* (Zygomycota: Entomophthorales) Scale bar = 0.3 mm.

Entomopathogenic fungi, which cause disease in insects, include at least 14 species that attack aphids. *Beauveria bassiana* is mass-produced and used to manage a wide variety of insect pests including whiteflies, thrips, aphids and weevils. *Lecanicillium* spp. are deployed against white flies, thrips and aphids. *Metarhizium* spp. are used against pests including beetles, locusts and other grasshoppers, Hemiptera, and spider mites. *Paecilomyces fumosoroseus* is effective against white flies, thrips and aphids; *Purpureocillium lilacinus* is used against root-knot nematodes, and 89 *Trichoderma* species against certain plant pathogens. *Trichoderma viride* has been used against Dutch elm disease, and has shown some effect in suppressing silver leaf, a disease of stone fruits caused by the pathogenic fungus *Chondrostereum purpureum*.

The fungi *Cordyceps* and *Metacordyceps* are deployed against a wide spectrum of arthropods. *Entomophaga* is effective against pests such as the green peach aphid.

Several members of Chytridiomycota and Blastocladiomycota have been explored as agents of biological control. From Chytridiomycota, *Synchytrium solstitiale* is being considered as a control agent of the yellow star thistle (*Centaurea solstitialis*) in the United States.

Viruses

Baculoviruses are specific to individual insect host species and have been shown to be useful in biological pest control. For example, the Lymantria dispar multicapsid nuclear polyhedrosis virus has been used to spray large areas of forest in North America where larvae of the gypsy moth are causing serious defoliation. The moth larvae are killed by the virus they have eaten and die, the disintegrating cadavers leaving virus particles on the foliage to infect other larvae.

A mammalian virus, the rabbit haemorrhagic disease virus has been introduced to Australia and to New Zealand to attempt to control the European rabbit populations there.

Algae

Lagenidium giganteum is a water-borne mould that parasitizes the larval stage of mosquitoes. When applied to water, the motile spores avoid unsuitable host species and search out suitable mosquito larval hosts. This alga has the advantages of a dormant phase, resistant to desiccation, with slow-release characteristics over several years. Unfortunately, it is susceptible to many chemicals used in mosquito abatement programmes.

Plants

The legume vine *Mucuna pruriens* is used in the countries of Benin and Vietnam as a biological control for problematic *Imperata cylindrica* grass. *Mucuna pruriens* is said not to be invasive outside its cultivated area. *Desmodium uncinatum* can be used in push-pull farming to stop the parasitic plant, *Striga*.

Other Methods

Combined use of Parasitoids and Pathogens

In cases of massive and severe infection of invasive pests, techniques of pest control are often used in combination. An example is the emerald ash borer, *Agrilus planipennis*, an invasive beetle from China, which has destroyed tens of millions of ash trees in its introduced range in North America. As part of the campaign against it, from 2003 American scientists and the Chinese Academy of Forestry searched for its natural enemies in the wild, leading to the discovery of several parasitoid wasps, namely *Tetrastichus planipennisi*, a gregarious larval endoparasitoid,*Oobius agrili*, a solitary, parthenogenic egg parasitoid, and *Spathius agrili*, a gregarious larval ectoparasitoid. These have been introduced and released into the United States of America as a possible biological control of the emerald ash borer. Initial results have shown promise with *Tetrastichus planipennisi* and it is now being released along with *Beauveria bassiana*, a fungal pathogen with known insecticidal properties.

Indirect Control

Pests may be controlled by biological control agents that do not prey directly upon them. For example, the Australian bush fly, *Musca vetustissima*, is a major nuisance pest in Australia, but native decomposers found in Australia are not adapted to feeding on cow dung, which is where bush flies breed. Therefore, the Australian Dung Beetle Project (1965–1985), led by Dr. George Bornemissza of the Commonwealth Scientific and Industrial Research Organisation, released forty-nine species of dung beetle, with the aim of reducing the amount of dung and therefore also the potential breeding sites of the fly.

Side-effects

Biological control can affect biodiversity through predation, parasitism, pathogenicity, competition, or other attacks on non-target species. An introduced control does not always target only the intended pest species; it can also target native species. In Hawaii during the 1940s parasitic wasps were introduced to control a lepidopteran pest and the wasps are still found there today. This may have a negative impact on the native ecosystem, however, host range and impacts need to be studied before declaring their impact on the environment.

Vertebrate animals tend to be generalist feeders, and seldom make good biological control agents; many of the classic cases of "biocontrol gone awry" involve vertebrates. For example, the cane toad (*Bufo marinus*) was intentionally introduced to Australia to control the greyback cane beetle (*Dermolepida albohirtum*), and other pests of sugar cane. 102 toads were obtained from Hawaii and bred in captivity to increase their numbers until they were released into the sugar cane fields of the tropic north in 1935. It was later discovered that the toads could not jump very high and so were unable to eat the cane beetles which stayed up on the upper stalks of the cane plants. However the toad thrived by feeding on other insects and it soon spread very rapidly; it took over native amphibian habitat and brought foreign disease to native toads and frogs, dramatically reducing their populations. Also when it is threatened or handled, the cane toad releases poison from parotoid glands on its shoulders; native Australian species such as goannas, tiger snakes, dingos and northern quolls that attempted to eat the toad were harmed or killed. However, there has been some recent evidence that native predators are adapting, both physiologically and through changing their behaviour, so in the long run, their populations may recover.

Rhinocyllus conicus, a seed-feeding weevil, was introduced to North America to control exotic musk thistle (*Carduus nutans*) and Canadian thistle (*Cirsium arvense*). However the weevil also attacks native thistles, harming such species as the endemic Platte thistle (*Cirsium neomexicanum*) by selecting larger plants (which reduced the gene pool), reducing seed production and ultimately threatening the species' survival.

The small Asian mongoose (*Herpestus javanicus*) was introduced to Hawaii in order to control the rat population. However it was diurnal and the rats emerged at night, and it preyed on the endemic birds of Hawaii, especially their eggs, more often than it ate the rats, and now both rats and mongooses threaten the birds. This introduction was undertaken without understanding the consequences of such an action. No regulations existed at the time, and more careful evaluation should prevent such releases now.

The sturdy and prolific eastern mosquitofish (*Gambusia holbrooki*) is a native of the southeastern United States and was introduced around the world in the 1930s and 40s to feed on mosquito larvae and thus combat malaria. However, it has thrived at the expense of local species, causing a decline of endemic fish and frogs through competition for food resources, as well as through eating their eggs and larvae. In Australia, the mosquitofish is the subject of discussion as to how best to control it; in 1989 it was said that "biological population control is well beyond present capabilities", and this remains the position.

Grower Education

A potential obstacle to the adoption of biological pest control measures is growers sticking to the familiar use of pesticides. It has been claimed that many of the pests that are controlled today using pesticides, actually became pests because pesticide use reduced or eliminated natural predators. A method of increasing grower adoption of biocontrol involves is letting growers learn by doing, for example showing them simple field experiments, having observations of live predation of pests, or collections of parasitised pests. In the Philippines, early season sprays against leaf folder caterpillars were common practice, but growers were asked to follow a 'rule of thumb' of not spraying against leaf folders for the first 30 days after transplanting; participation in this resulted in a reduction of insecticide use by 1/3 and a change in grower perception of insecticide use.

Companion Planting

Companion planting in gardening and agriculture is the planting of different crops in proximity for pest control, pollination, providing habitat for beneficial creatures, maximizing use of space, and to otherwise increase crop productivity. Companion planting is a form of polyculture.

Companion planting is used by farmers and gardeners in both industrialized and developing countries for many reasons. Many of the modern principles of companion planting were present many centuries ago in cottage gardens in England and forest gardens in Asia, and thousands of years ago in Mesoamerica.

History

In China, mosquito ferns (*Azolla* spp.) have been used for at least a thousand years as companion plants for rice crops. They host a cyanobacterium that fixes nitrogen from the atmosphere, and they block light from plants that would compete with the rice.

Companion planting was practiced in various forms by the indigenous peoples of the Americas prior to the arrival of Europeans. These peoples domesticated squash 8,000 to 10,000 years ago, then maize, then common beans, forming the Three Sisters agricultural technique. The cornstalk served as a trellis for the beans to climb, and the beans fixed nitrogen, benefitting the maize.

Companion planting was widely promoted in the 1970s as part of the organic gardening movement. It was encouraged for pragmatic reasons, such as natural trellising, but mainly with the idea that different species of plant may thrive more when close together. It is also a technique frequently used in permaculture, together with mulching, polyculture, and changing of crops.

Examples of Companion Plants

Nasturtium (*Tropaeolum majus*) is a food plant of some caterpillars which feed primarily on members of the cabbage family (brassicas), and some gardeners claim that planting them around brassicas protects the food crops from damage, as eggs of the pests are preferentially laid on the nasturtium. This practice is called trap cropping.

The smell of the foliage of marigolds is claimed to deter aphids from feeding on neighbouring crops. Marigolds with simple flowers also attract nectar-feeding adult hover-flies, the larvae of which are predators of aphids.

Various legume crops benefit from being commingled with a grassy nurse crop. For example, common vetch or hairy vetch is planted together with rye or winter wheat to make a good cover crop or green manure (or both).

Versions

There are a number of systems and ideas using companion planting.

Square foot gardening attempts to protect plants from many normal gardening problems by packing them as closely together as possible, which is facilitated by using companion plants, which can be closer together than normal.

Another system using companion planting is the forest garden, where companion plants are intermingled to create an actual ecosystem, emulating the interaction of up to seven levels of plants in a forest or woodland.

Organic gardening may make use of companion planting, since many synthetic means of fertilizing, weed reduction and pest control are forbidden.

Host-finding Disruption

Recent studies on host-plant finding have shown that flying pests are far less successful if their host-plants are surrounded by any other plant or even "decoy-plants" made of green plastic, cardboard, or any other green material.

The host-plant finding process occurs in phases:

The first phase is stimulation by odours characteristic to the host-plant. This induces the insect to try to land on the plant it seeks. But insects avoid landing on brown (bare) soil. So if only the host-plant is present, the insects will quasi-systematically find it by simply landing on the only green thing around. This is called (from the point of view of the insect) "appropriate landing". When it does an "inappropriate landing", it flies off to any other nearby patch of green. It eventually leaves the area if there are too many 'inappropriate' landings.

The second phase of host-plant finding is for the insect to make short flights from leaf to leaf to assess the plant's overall suitability. The number of leaf-to-leaf flights varies according to the insect species and to the host-plant stimulus received from each leaf. The insect must accumulate sufficient stimuli from the host-plant to lay eggs; so it must make a certain number of consecutive 'appropriate' landings. Hence if it makes an 'inappropriate landing', the assessment of that plant is negative, and the insect must start the process anew.

Thus it was shown that clover used as a ground cover had the same disruptive effect on eight pest species from four different insect orders. An experiment showed that 36% of cabbage root flies laid eggs beside cabbages growing in bare soil (which resulted in no crop), compared to only 7% beside cabbages growing in clover (which allowed a good crop). Simple decoys made of green cardboard also disrupted appropriate landings just as well as did the live ground cover.

Companion Plant Categories

The use of companion planting can be of benefit to the grower in a number of different ways, including:

- Hedged investment – the growing of different crops in the same space increases the odds of some yield being given, even if one crop fails.

- Increased level interaction – when crops are grown on different levels in the same space, such as providing ground cover or one crop working as a trellis for another, the overall yield of a plot may be increased.

- Protective shelter is when one type of plant may serve as a wind break or provide shade for another.

- Pest suppression – some companion plants may help prevent pest insects or pathogenic fungi from damaging the crop, through chemical means.

- Predator recruitment and positive hosting – The use of companion plants that produce copious nectar or pollen in a vegetable garden (insectary plants) may help encourage higher populations of beneficial insects that control pests, as some beneficial predatory insects only consume pests in their larval form and are nectar or pollen feeders in their adult form.

- Trap cropping – some companion plants are claimed to attract pests away from others.

- Pattern disruption – in a monoculture pests spread easily from one crop plant to the next, whereas such easy progress may be disrupted by surrounding companion plants of a different type.

Crop Rotation

Crop rotation is the practice of growing a series of dissimilar or different types of crops in the same area in sequenced seasons.It is done so that the soil of farms is not used to only one type of nutrient. It helps in reducing soil erosion and increases soil fertility and crop yield.

Growing the same crop in the same place for many years in a row disproportionately depletes the soil of certain nutrients. With rotation, a crop that leaches the soil of one kind of nutrient is followed during the next growing season by a dissimilar crop that returns that nutrient to the soil or draws a different ratio of nutrients. In addition, crop rotation mitigates the buildup of pathogens and pests that often occurs when one species is continuously cropped, and can also improve soil structure and fertility by increasing biomass from varied root structures.

Crop rotation is used in both conventional and organic farming systems.

History

It has long been recognized that suitable rotations – such as planting spring crops for livestock in place of grains for human consumption – make it possible to restore or to maintain a productive soil. Middle Eastern farmers practiced crop rotation in 6000 BC without understanding the chemistry, alternately planting legumes and cereals. In the Bible chapter of Leviticus 25, God instructs the Israelites to observe a 'Sabbath of the Land'. Every seventh year they would not till, prune or even control insects. The Roman writer, Cato the Elder, recommended that farmers "save carefully goat, sheep, cattle, and all other dung". In Europe, since the times of Charlemagne, there was a transition from a two-field crop rotation to a three-field crop rotation. Under a two-field rotation, half the land was planted in a year, while the other half lay fallow. Then, in the next year, the two fields were reversed.

From the end of the Middle Ages until the 20th century, three-year rotation was practiced by farmers in Europe. Under three-field rotation, the land was divided into three parts. One section was planted in the autumn with rye or winter wheat, followed by spring oats or barley, or other crops such as peas, lentils, or beans and the third field was left fallow. The three fields were rotated in this manner so that every three years, a field would rest and be fallow. Under the two-field system, if one has a total of 600 acres (2.4 km^2) of fertile land, one would only plant 300 acres. Under the new three-field rotation system, one would plant (and therefore harvest) 400 acres. But, the additional crops had a more significant effect than mere productivity. Since the spring crops were mostly legumes, they increased the overall nutrition of the people of Northern Europe.

A four-field rotation was pioneered by farmers, namely in the region Waasland in the early 16th century and popularised by the British agriculturist Charles Townshend in the 18th century. The system (wheat, turnips, barley and clover), opened up a fodder crop and grazing crop allowing livestock to be bred year-round. The four-field crop rotation was a key development in the British Agricultural Revolution.

George Washington Carver studied crop rotation methods in the United States, teaching southern farmers to rotate soil-depleting crops like cotton with soil-enriching crops like peanuts and peas.

In the Green Revolution, the traditional practice of crop rotation gave way in some parts of the world to the practice of supplementing the chemical inputs to the soil through top dressing with fertilizers, e.g. adding ammonium nitrate or urea and restoring soil pH with lime in the search for increased yields, preparing soil for specialist crops, and seeking to reduce waste and inefficiency by simplifying planting and harvesting.

Crop Choice

A preliminary assessment of crop interrelationships can be found in how each crop: (1) contributes to soil organic matter (SOM) content, (2) provides for pest management, (3) manages deficient or excess nutrients, and (4) how it contributes to or controls for soil erosion.

Crop choice is often a related to the goal the farmer is looking to achieve with the rotation, which could be weed management, increasing available nitrogen in the soil, controlling for erosion, or increasing soil structure and biomass, to name a few. When discussing crop rotations, crops are classified in different ways depending on what quality is being assessed: by family, by nutrient needs/benefits, and/ or by profitability (i.e. cash crop versus cover crop). For example, giving adequate attention to plant family is essential to mitigating pests and pathogens. However, many farmers have success managing rotations by planning sequencing and cover crops around desirable cash crops. The following is a simplified classification based on crop quality and purpose.

Row Crops

Many crops which are critical for the market, like vegetables, are row crops (that is, grown in tight rows). While often the most profitable for farmers, these crops are more taxing on the soil Row crops typically have low biomass and shallow roots: this means the plant contributes low residue to the surrounding soil and has limited effects on structure. With much of the soil around the plant is exposed to disruption by rainfall and traffic, fields with row crops experience faster break down of organic matter by microbes, leaving fewer nutrients for future plants.

In short, while these crops may be profitable for the farm, they are nutrient depleting. Crop rotation practices exist to strike a balance between short-term profitability and long-term productivity.

Legumes

A great advantage of crop rotation comes from the interrelationship of nitrogen fixing-crops with nitrogen demanding crops. Legumes, like alfalfa and clover, collect available nitrogen from the soil in nodules on their root structure. When the plant is harvested, the biomass of uncollected roots breaks down, making the stored nitrogen available to future crops. Legumes are also a valued green manure: a crop that collects nutrients and fixes them at soil depths accessible to future crops.

In addition, legumes have heavy tap roots that burrow deep into the ground, lifting soil for better tilth and absorption of water.

Grasses and Cereals

Cereal and grasses are frequent cover crops because of the many advantages they supply to soil quality and structure. The dense and far-reaching root systems give ample structure to surrounding soil and provide significant biomass for soil organic matter.

Grasses and cereals are key in weed management as they compete with undesired plants for soil space and nutrients.

Green Manure

Green manure is a crop that is mixed into the soil. Both nitrogen-fixing legumes and nutrient scavengers, like grasses, can be used as green manure. Green manure of legumes is an excellent source of nitrogen, especially for organic systems, however, legume biomass doesn't contribute to lasting soil organic matter like grasses do.

Planning a Rotation

There are numerous factors that must be taken into consideration when planning a crop rotation. Planning an effective rotation requires weighing fixed and fluctuating production circumstances, including, but not limited to: market, farm size, labor supply, climate, soil type, growing practices, etc. Moreover, a crop rotation must consider in what condition one crop will leave the soil for the succeeding crop and how one crop can be seeded with another crop. For example, a nitrogen-fixing crop, like a legume, should always proceed a nitrogen depleting one; similarly, a low residue crop (i.e. a crop with low biomass) should be offset with a high biomass cover crop, like a mixture of grasses and legumes.

There is no limit to the number of crops that can be used in a rotation, or the amount of time a rotation takes to complete. Decisions about rotations are made years prior, seasons prior, or even at the very last minute when an opportunity to increase profits or soil quality presents itself. In short, there is no singular formula for rotation, but many considerations to take into account.

Implementation

Crop rotation systems may be enriched by the influences of other practices such as the addition of livestock and manure, intercropping or multiple cropping, and organic management low in pesticides and synthetic fertilizers.

Incorporation of Livestock

Introducing livestock makes the most efficient use of critical sod and cover crops; livestock (through manure) are able to distribute the nutrients in these crops through-out the soil rather than removing nutrients from the farm through the sale of hay. In systems where use of farm livestock would violate reservations growers or consumers may have about animal exploitation, efforts are made to surrogate this input through livestock in the soil, namely worms and microorganisms.

In Sub-Saharan Africa, as animal husbandry becomes less of a nomadic practice many herders have begun integrating crop production into their practice. This is known as mixed farming, or the practice of crop cultivation with the incorporation of raising cat-tle, sheep and/or goats by the same economic entity, is increasingly common. This in-teraction between the animal, the land and the crops are being done on a small scale all across this region. Crop residues provide animal feed, while the animals provide manure for replenishing crop nutrients and draft power. Both processes are extremely important in this region of the world as it is expensive and logistically unfeasible to transport in synthetic fertilizers and large-scale machinery. As an additional benefit, the cattle, sheep and/or goat provide milk and can act as a cash crop in the times of economic hardship.

Organic Farming

Crop rotation is a required practice in order for a farm to receive organic certification in the United States. The "Crop Rotation Practice Standard" for the National Organic Program under the U.S. Code of Federal Regulations, section §205.205, states that:

Farmers are required to implement a crop rotation that maintains or builds soil organic matter, works to control pests, manages and conserves nutrients, and protects against erosion. Producers of perennial crops that aren't rotated may utilize other practices, such as cover crops, to maintain soil health.

In addition to lowering the need for inputs by controlling for pests and weeds and in-creasing available nutrients, crop rotation helps organic growers increase the amount of biodiversity on their farms. Biodiversity is also a requirement of organic certifica-tion, however, there are no rules in place to regulate or reinforce this standard. In-creasing the biodiversity of crops has beneficial effects on the surrounding ecosystem and can host a greater diversity of fauna, insects, and beneficial microorganism in the soil.< Some studies point to increased nutrient availability from crop rotation under organic systems compared to conventional practices as organic practices are less likely to inhibit of beneficial microbes in soil organic matter.

While multiple cropping and intercropping benefit from many of the same principals as crop rotation, they do not satisfy the requirement under the NOP.

Intercropping

Multiple cropping systems, such as intercropping or companion planting, offer more diversity and complexity within the same season or rotation, for example the three sisters. An example of companion planting is the inter-planting of corn with pole beans and vining squash or pumpkins. In this system, the beans provide nitrogen; the corn provides support for the beans and a "screen" against squash vine borer; the vining squash provides a weed suppressive canopy and discourages corn-hungry raccoons.

Double-cropping is common where two crops, typically of different species, are grown sequentially in the same growing season, or where one crop (e.g. vegetable) is grown continuously with a cover crop (e.g. wheat). This is advantageous for small farms, who often cannot afford to leave cover crops to replenish the soil for extended periods of time, as larger farms can. When multiple cropping is implemented on small farms, these systems can maximize benefits of crop rotation on available land resources.

Benefits

Agronomists describe the benefits to yield in rotated crops as "The Rotation Effect". There are many found benefits of rotation systems: however, there is no specific scientific basis for the sometimes 10-25% yield increase in a crop grown in rotation versus monoculture. The factors related to the increase are simply described as alleviation of the negative factors of monoculture cropping systems. Explanations due to improved nutrition; pest, pathogen, and weed stress reduction; and improved soil structure have been found in some cases to be correlated, but causation has not been determined for the majority of cropping systems.

Other benefits of rotation cropping systems include production cost advantages. Overall financial risks are more widely distributed over more diverse production of crops and/or livestock. Less reliance is placed on purchased inputs and over time crops can maintain production goals with fewer inputs. This in tandem with greater short and long term yields makes rotation a powerful tool for improving agricultural systems.

Soil Organic Matter

The use of different species in rotation allows for increased soil organic matter (SOM), greater soil structure, and improvement of the chemical and biological soil environment for crops. With more SOM, water infiltration and retention improves, providing increased drought tolerance and decreased erosion.

Soil organic matter is a mix of decaying material from biomass with active microorganisms. Crop rotation, by nature, increases exposure to biomass from sod, green manure, and a various other plant debris. The reduced need for intensive tillage under crop rotation allows biomass aggregation to lead to greater nutrient retention and utilization, decreasing the need for added nutrients. With tillage, disruption and oxidation of soil creates a less conducive environment for diversity and proliferation of microorganisms in the soil. These microorganisms are what make nutrients available to plants. So, where "active" soil organic matter is a key to productive soil, soil with low microbial activity provides significantly fewer nutrients to plants; this is true even though the quantity of biomass left in the soil may be the same.

Soil microorganisms also decrease pathogen and pest activity through competition. In addition, plants produce root exudates and other chemicals which manipulate their soil environment as well as their weed environment. Thus rotation allows increased yields from nutrient availability but also alleviation of allelopathy and competitive weed environments.

Carbon Sequestration

Studies have shown that crop rotations greatly increase soil organic carbon (SOC) content, the main constituent of soil organic matter. Carbon, along with hydrogen and oxygen, is a macronutrient for plants. Highly diverse rotations spanning long periods of time have shown to be even more effective in increasing SOC, while soil disturbances (e.g. from tillage) are responsible for exponential decline in SOC levels. In Brazil, conservation to no-till methods combined with intensive crop rotations has been shown an SOC sequestration rate of 0.41 tonnes per hectare per year.

In addition to enhancing crop productivity, sequestration of atmospheric carbon has great implications in reducing rates of climate change by removing carbon dioxide from the air.

Nitrogen Fixing

Rotating crops adds nutrients to the soil. Legumes, plants of the family Fabaceae, for instance, have nodules on their roots which contain nitrogen-fixing bacteria called rhizobia. It therefore makes good sense agriculturally to alternate them with cereals (family Poaceae) and other plants that require nitrates.

Pathogen and Pest Control

Crop rotation is also used to control pests and diseases that can become established in the soil over time. The changing of crops in a sequence decreases the population level of pests by (1) interrupting pest life cycles and (2) interrupting pest habitat. Plants within the same taxonomic family tend to have similar pests and pathogens. By regularly changing crops and keeping the soil occupied by cover crops instead of lying fallow, pest cycles can be broken or limited, especially cycles that benefit from overwintering in residue. For example, root-knot nematode is a serious problem for some plants in warm climates and sandy soils, where it slowly builds up to high levels in the soil, and can severely damage plant productivity by cutting off circulation from the plant roots. Growing a crop that is not a host for root-knot nematode for one season greatly reduces the level of the nematode in the soil, thus making it possible to grow a susceptible crop the following season without needing soil fumigation.

This principle is of particular use in organic farming, where pest control must be achieved without synthetic pesticides.

Weed Management

Integrating certain crops, especially cover crops, into crop rotations is of particular value to weed management. These crops crowd out weed through competition. In addition, the sod and compost from cover crops and green manure slows the growth of what weeds are still able to make it through the soil, giving the crops further competitive advantage. By removing slowing the growth and proliferation of weeds while cover crops are cultivated, farmers greatly reduce the presence of weeds for future crops, including shallow rooted and row crops, which are less resistant to weeds. Cover crops are, therefore, considered conservation crops because they protect otherwise fallow land from becoming overrun with weeds.

This system has advantages over other common practices for weeds management, such as tillage. Tillage is meant to inhibit growth of weeds by overturning the soil; however, this has a countering effect of exposing weed seeds that may have gotten buried and burying valuable crop seeds. Under crop rotation, the number of viable seeds in the soil is reduced through the reduction of the weed population.

Preventing Soil Erosion

Crop rotation can significantly reduce the amount of soil lost from erosion by water. In areas that are highly susceptible to erosion, farm management practices such as zero and reduced tillage can be supplemented with specific crop rotation methods to reduce raindrop impact, sediment detachment, sediment transport, surface runoff, and soil loss.

Protection against soil loss is maximized with rotation methods that leave the greatest mass of crop stubble (plant residue left after harvest) on top of the soil. Stubble cover in contact with the soil minimizes erosion from water by reducing overland flow velocity, stream power, and thus the ability of the water to detach and transport sediment. Soil Erosion and Cill prevent the disruption and detachment of soil aggregates that cause macropores to block, infiltration to decline, and runoff to increase. This significantly improves the resilience of soils when subjected to periods of erosion and stress.

The effect of crop rotation on erosion control varies by climate. In regions under relatively consistent climate conditions, where annual rainfall and temperature levels are assumed, rigid crop rotations can produce sufficient plant growth and soil cover. In regions where climate conditions are less predictable, and unexpected periods of rain and drought may occur, a more flexible approach for soil cover by crop rotation is necessary. An opportunity cropping system promotes adequate soil cover under these erratic climate conditions. In an opportunity cropping system, crops are grown when soil water is adequate and there is a reliable sowing window. This form of cropping system is likely to produce better soil cover than a rigid crop rotation because crops are only sown under optimal conditions, whereas rigid systems are not necessarily sown in the best conditions available.

Crop rotations also affect the timing and length of when a field is subject to fallow. This is very important because depending on a particular region's climate, a field could be the most vulnerable to erosion when it is under fallow. Efficient fallow management is an essential part of reducing erosion in a crop rotation system. Zero tillage is a fundamental management practice that promotes crop stubble retention under longer unplanned fallows when crops cannot be planted. Such management practices that succeed in retaining suitable soil cover in areas under fallow will ultimately reduce soil loss.

Biodiversity

Increasing the biodiversity of crops has beneficial effects on the surrounding ecosystem and can host a greater diversity of fauna, insects, and beneficial microorganisms in the soil. Some studies point to increased nutrient availability from crop rotation under organic systems compared to conventional practices as organic practices are less likely to inhibit of beneficial microbes in soil organic matter, such as arbuscular mycorrhizae, which increase nutrient uptake in plants. Increasing biodiversity also increases the resilience of agro-ecological systems.

Farm Productivity

Crop rotation contributes to increased yields through improved soil nutrition. By requiring planting and harvesting of different crops at different times, more land can be farmed with the same amount of machinery and labour.

Risk Management

Different crops can reduce the risks of adverse weather for the individual farmer.

Challenges

While crop rotation requires a great deal of planning, crop choice must respond to a number of fixed conditions (soil type, topography, climate, and irrigation) in addition to conditions that may change dramatically from year to the next (weather, market, labor supply). In this way, it is unwise to plan to crops years in advance. Improper implementation of a crop rotation plan may lead to imbalances in the soil nutrient composition or a buildup of pathogens affecting a critical crop. The consequences of faulty rotation may take years to become apparent even to experienced soil scientists and can take just as long to correct.

Many challenges exist within the practices associated with crop rotation. For example, green manure from legumes can lead to an invasion of snails or slugs and the decay from green manure can occasionally suppress the growth of other crops.

Integrated Pest Management

Integrated pest management (IPM), also known as integrated pest control (IPC) is a broad-based approach that integrates practices for economic control of pests. IPM aims to suppress pest populations below the economic injury level (EIL). The UN's Food and Agriculture Organisation defines IPM as "the careful consideration of all available pest control techniques and subsequent integration of appropriate measures that discourage the development of pest populations and keep pesticides and other interventions to levels that are economically justified and reduce or minimize risks to human health and the environment. IPM emphasizes the growth of a healthy crop with the least possible disruption to agro-ecosystems and encourages natural pest control mechanisms." Entomologists and ecologists have urged the adoption of IPM pest control since the 1970s. IPM allows for safer pest control. This includes managing insects, plant pathogens and weeds.

Globalization and increased mobility often allow increasing numbers of invasive species to cross national borders. IPM poses the least risks while maximizing benefits and reducing costs.

For their leadership in developing and spreading IPM worldwide, Perry Adkisson and Ray F. Smith received the 1997 World Food Prize.

History

Shortly after World War II, when synthetic insecticides became widely available, entomologists in California developed the concept of "supervised insect control". Around the same time, entomologists in the US Cotton Belt were advocating a similar approach. Under this scheme, insect control was "supervised" by qualified entomologists and insecticide applications were based on conclusions reached from periodic monitoring of pest and natural-enemy populations. This was viewed as an alternative to calendar-based programs. Supervised control was based on knowledge of the ecology and analysis of projected trends in pest and natural-enemy populations.

Supervised control formed much of the conceptual basis for the "integrated control" that University of California entomologists articulated in the 1950s. Integrated control sought to identify the best mix of chemical and biological controls for a given insect pest. Chemical insecticides were to be used in the manner least disruptive to biological control. The term "integrated" was thus synonymous with "compatible." Chemical controls were to be applied only after regular monitoring indicated that a pest population had reached a level (the economic threshold) that required treatment to prevent the population from reaching a level (the economic injury level) at which economic losses would exceed the cost of the control measures.

IPM extended the concept of integrated control to all classes of pests and was expanded to include all tactics. Controls such as pesticides were to be applied as in integrated control, but these now had to be compatible with tactics for all classes of pests. Other tactics, such as host-plant resistance and cultural manipulations, became part of the IPM framework. IPM combined entomologists, plant pathologists, nematologists and weed scientists.

In the United States, IPM was formulated into national policy in February 1972 when President Richard Nixon directed federal agencies to take steps to advance the application of IPM in all relevant sectors. In 1979, President Jimmy Carter established an interagency IPM Coordinating Committee to ensure development and implementation of IPM practices.

Applications

IPM is used in agriculture, horticulture, human habitations, preventive conservation and general pest control, including structural pest management, turf pest management and ornamental pest management.

Principles

An American IPM system is designed around six basic components:

Acceptable pest levels—The emphasis is on *control*, not *eradication*. IPM holds that wiping out an entire pest population is often impossible, and the attempt can be expensive and unsafe. IPM programmes first work to establish acceptable pest levels, called action thresholds, and apply controls if those thresholds are crossed. These thresholds are pest and site specific, meaning that it may be acceptable at one site to have a weed such as white clover, but not at another site. Allowing a pest population to survive at a reasonable threshold reduces selection pressure. This lowers the rate at which a pest develops resistance to a control, because if almost all pests are killed then those that have resistance will provide the genetic basis of the future population. Retaining a significant number unresistant specimens dilutes the prevalence of any resistant genes that appear. Similarly, the repeated use of a single class of controls will create pest populations that are more resistant to that class, whereas alternating among classes helps prevent this.

Preventive cultural practices—Selecting varieties best for local growing conditions and maintaining healthy crops is the first line of defense. Plant quarantine and 'cultural techniques' such as crop sanitation are next, e.g., removal of diseased plants, and cleaning pruning shears to prevent spread of infections. Beneficial fungi and bacteria are added to the potting media of horticultural crops vulnerable to root diseases, greatly reducing the need for fungicides.

Monitoring—Regular observation is critically important. Observation is broken into in-spection and identification. Visual inspection, insect and spore traps, and other meth-ods are used to monitor pest levels. Record-keeping is essential, as is a thorough knowl-edge target pest behavior and reproductive cycles. Since insects are cold-blooded, their physical development is dependent on area temperatures. Many insects have had their development cycles modeled in terms of degree-days. The degree days of an environ-ment determines the optimal time for a specific insect outbreak. Plant pathogens follow similar patterns of response to weather and season.

Mechanical controls—Should a pest reach an unacceptable level, mechanical methods are the first options. They include simple hand-picking, barriers, traps, vacuuming and tillage to disrupt breeding.

Biological controls—Natural biological processes and materials can provide control, with acceptable environmental impact, and often at lower cost. The main approach is to promote beneficial insects that eat or parasitize target pests. Biological insecticides, derived from naturally occurring microorganisms (*e.g.*—Bt, entomopathogenic fungi and entomopathogenic nematodes), also fall in this category. Further 'biology-based' or 'ecological' techniques are under evaluation.

Responsible use—Synthetic pesticides are used as required and often only at specific times in a pest's life cycle. Many newer pesticides are derived from plants or natural-ly occurring substances (*e.g.*—nicotine, pyrethrum and insect juvenile hormone ana-logues), but the toxophore or active component may be altered to provide increased biological activity or stability. Applications of pesticides must reach their intended targets. Matching the application technique to the crop, the pest, and the pesticide is critical. The use of low-volume spray equipment reduces overall pesticide use and labor cost.

An IPM regime can be simple or sophisticated. Historically, the main focus of IPM programmes was on agricultural insect pests. Although originally developed for agri-cultural pest management, IPM programmes are now developed to encompass diseas-es, weeds and other pests that interfere with management objectives for sites such as residential and commercial structures, lawn and turf areas, and home and community gardens.

Process

IPM is the selection and use of pest control actions that will ensure favourable econom-ic, ecological and social consequences and is applicable to most agricultural, public health and amenity pest management situations. The IPM process starts with moni-toring, which includes inspection and identification, followed by the establishment of economic injury levels. The economic injury levels set the economic threshold level. That is the point when pest damage (and the benefits of treating the pest) exceed the cost of treatment. This can also be an action threshold level for determining an unacceptable level that is not tied to economic injury. Action thresholds are more common in structural pest management and economic injury levels in classic agricultural pest management. An example of an action threshold is one fly in a hospital operating room is not acceptable, but one fly in a pet kennel would be acceptable. Once a threshold has been crossed by the pest population action steps need to be taken to reduce and control the pest. Integrated pest management employ a variety of actions including cultural controls, including physical barriers, biological controls, including adding and conserving natural predators and enemies to the pest, and finally chemical controls or pesticides. Reliance on knowledge, experience, observation and integration of multiple techniques makes IPM appropriate for organic farming (excluding synthetic pesticides). These may or may not include materials listed on the Organic Materials Review Institute (OMRI) Although the pesticides and particularly insecticides used in organic farming and organic gardening are generally safer than synthetic pesticides, they are not always more safe or environmentally friendly than synthetic pesticides and can cause harm. For conventional farms IPM can reduce human and environmental exposure to hazardous chemicals, and potentially lower overall costs.

Risk assessment usually includes four issues: 1) characterization of biological control agents, 2) health risks, 3) environmental risks and 4) efficacy.

Mistaken identification of a pest may result in ineffective actions. E.g., plant damage due to over-watering could be mistaken for fungal infection, since many fungal and viral infections arise under moist conditions.

Monitoring begins immediately, before the pest's activity becomes significant. Monitoring of agricultural pests includes tracking soil/planting media fertility and water quality. Overall plant health and resistance to pests is greatly influenced by pH, alkalinity, of dissolved mineral and Oxygen Reduction Potential. Many diseases are waterborne, spread directly by irrigation water and indirectly by splashing.

Once the pest is known, knowledge of its lifecycle provides the optimal intervention points. For example, weeds reproducing from last year's seed can be prevented with mulches and pre-emergent herbicide.

Pest-tolerant crops such as soybeans may not warrant interventions unless the pests are numerous or rapidly increasing. Intervention is warranted if the expected cost of damage by the pest is more than the cost of control. Health hazards may require intervention that is not warranted by economic considerations.

Specific sites may also have varying requirements. E.g., white clover may be acceptable on the sides of a tee box on a golf course, but unacceptable in the fairway where it could confuse the field of play.

Possible interventions include mechanical/physical, cultural, biological and chemical.

Mechanical/physical controls include picking pests off plants, or using netting or other material to exclude pests such as birds from grapes or rodents from structures. Cultural controls include keeping an area free of conducive conditions by removing waste or diseased plants, flooding, sanding, and the use of disease-resistant crop varieties. Biological controls are numerous. They include: conservation of natural predators or augmentation of natural predators, Sterile insect technique (SIT).

Augmentation, inoculative release and inundative release are different methods of biological control that affect the target pest in different ways. Augmentative control includes the periodic introduction of predators. With inundative release, predators are collected, mass-reared and periodically released in large numbers into the pest area. This is used for an immediate reduction in host populations, generally for annual crops, but is not suitable for long run use. With inoculative release a limited number of beneficial organisms are introduced at the start of the growing season. This strategy offers long term control as the organism's progeny affect pest populations throughout the season and is common in orchards. With seasonal inoculative release the beneficials are collected, mass-reared and released seasonally to maintain the beneficial population. This is commonly used in greenhouses. In America and other western countries, inundative releases are predominant, while Asia and the eastern Europe more commonly use inoculation and occasional introductions.

The Sterile insect technique (SIT) is an Area-Wide IPM program that introduces sterile male pests into the pest population to trick females into (unsuccessful) breeding encounters, providing a form of birth control and reducing reproduction rates. The biological controls mentioned above only appropriate in extreme cases, because in the introduction of new species, or supplementation of naturally occurring species can have detrimental ecosystem effects. Biological controls can be used to stop invasive species or pests, but they can become an introduction path for new pests.

Chemical controls include horticultural oils or the application of insecticides and herbicides. A Green Pest Management IPM program uses pesticides derived from plants, such as botanicals, or other naturally occurring materials.

Pesticides can be classified by their modes of action. Rotating among materials with different modes of action minimizes pest resistance.

Evaluation is the process of assessing whether the intervention was effective, whether it produced unacceptable side effects, whether to continue, revise or abandon the program.

Southeast Asia

The Green Revolution of the 1960s and '70s introduced sturdier plants that could support the heavier grain loads resulting from intensive fertilizer use. Pesticide imports by 11 Southeast Asian countries grew nearly sevenfold in value between 1990 and 2010, according to FAO statistics, with disastrous results. Rice farmers become accustomed to spraying soon after planting, triggered by signs of the leaf folder moth, which appears early in the growing season. It causes only superficial damage and doesn't reduce yields. In 1986, Indonesia banned 57 pesticides and completely stopped subsidizing their use. Progress was reversed in the 2000s, when growing production capacity, particularly in China, reduced prices. Rice production in Asia more than doubled. But it left farmers believing more is better—whether it's seed, fertilizer, or pesticides.

The brown planthopper (Nilaparvata lugens), the farmers' main target, has become increasingly resistant. Since 2008, outbreaks have devastated rice harvests throughout Asia, but not in the Mekong Delta. Reduced spraying allowed natural predators to neutralize planthoppers in Vietnam. In 2010 and 2011, massive planthopper outbreaks hit 400,000 hectares of Thai rice fields, causing losses of about $64 million. The Thai government is now pushing the "no spray in the first 40 days" approach.

By contrast early spraying kills frogs, spiders, wasps and dragonflies that prey on the later-arriving and dangerous planthopper and produced resistant strains. Planthoppers now require pesticide doses 500 times greater than originally. Overuse indiscriminately kills beneficial insects and decimates bird and amphibian populations. Pesticides are suspected of harming human health and became a common means for rural Asians to commit suicide.

In 2001, scientists challenged 950 Vietnamese farmers to try IPM. In one plot, each farmer grew rice using their usual amounts of seed and fertilizer, applying pesticide as they chose. In a nearby plot, less seed and fertilizer were used and no pesticides were applied for 40 days after planting. Yields from the experimental plots was as good or better and costs were lower, generating 8% to 10% more net income. The experiment led to the "three reductions, three gains" campaign, claiming that cutting the use of seed, fertilizer and pesticide would boost yield, quality and income. Posters, leaflets, TV commercials and a 2004 radio soap opera that featured a rice farmer who gradually accepted the changes. It didn't hurt that a 2006 planthopper outbreak hit farmers using insecticides harder than those who didn't. Mekong Delta farmers cut insecticide spraying from five times per crop cycle to zero to one.

The Plant Protection Center and the International Rice Research Institute (IRRI) have been encouraging farmers to grow flowers, okra and beans on rice paddy banks, instead of stripping vegetation, as was typical. The plants attract bees and a tiny wasp that eats planthopper eggs, while the vegetables diversify farm incomes.

Agriculture companies offer bundles of pesticides with seeds and fertilizer, with incentives for volume purchases. A proposed law in Vietnam requires licensing pesticide dealers and government approval of advertisements to prevent exaggerated claims. Insecticides that target other pests, such as Scirpophaga incertulas (stem borer), the larvae of moth species that feed on rice plants allegedly yield gains of 21% with proper use.

References

- Baur, Fred. Insect Management for Food Storage and Processing. American Ass. of Cereal Chemists. pp. 162–165. ISBN 0-913250-38-4.

- Flint, Maria Louise & Dreistadt, Steve H. (1998). Clark, Jack K., ed. Natural Enemies Handbook: The Illustrated Guide to Biological Pest Control. University of California Press. ISBN 978-0-520-21801-7.

- Acorn, John (2007). Ladybugs of Alberta: Finding the Spots and Connecting the Dots. University of Alberta. p. 15. ISBN 978-0-88864-381-0.

- Johnson, Sue Ellen; Charles L. Mohler, (2009). Crop Rotation on Organic Farms: A Planning Manual, NRAES 177. Ithica, NY: National Resource, Agriculture, and Engineering Services (NRAES). ISBN 978-1-933395-21-0.

- Charles Perrings; Mark Herbert Williamson; Silvana Dalmazzone (1 January 2000). The Economics of Biological Invasions. Edward Elgar Publishing. ISBN 978-1-84064-378-7.

- Consoli, Fernando L.; Parra, José Roberto Postali; Trichogramma, Roberto Antônio Zucchi (28 September 2010). Egg Parasitoids in Agroecosystems with Emphasis on. Springer. ISBN 978-1-4020-9110-0.

- Rajeev K. Upadhyay; K.G. Mukerji; B. P. Chamola (30 November 2001). Biocontrol Potential and its Exploitation in Sustainable Agriculture: Volume 2: Insect Pests. Springer. pp. 261–. ISBN 978-0-306-46587-1.

- J. C. van Lenteren (2003). Quality Control and Production of Biological Control Agents: Theory and Testing Procedures. CABI. ISBN 978-0-85199-836-7.

- Kaya, Harry K. et al. (1993). "An Overview of Insect-Parasitic and Entomopathogenic Nematodes". In Bedding, R.A. Nematodes and the Biological Control of Insect Pests. CSIRO Publishing. ISBN 978-0-643-10591-1.

- Capinera, John L. (October 2005). "Featured creatures:". University of Florida website - Department of Entomology and Nematology. University of Florida. Retrieved 7 June 2016.

- "Moving on from the mongoose: the success of biological control in Hawai'i". Kia'i Moku. MISC. 18 April 2012. Retrieved 2 July 2016.

- Dufour, Rex (July 2015). Tipsheet: Crop Rotation in Organic Farming Systems (Report). National Center for Appropriate Technology. Retrieved May 4, 2016.

- Baldwin, Keith R. (June 2006). Crop Rotations on Organic Farms (PDF) (Report). Center for Environmental Farming Systems. Retrieved May 4, 2016.

- Coleman, Pamela (November 2012). Guide for Organic Crop Producers (PDF) (Report). National Organic Program. Retrieved May 4, 2016.

- Gegner, Lance; George Kuepper (August 2004). "Organic Crop Production Overview". National Center for Appropriate Technology. Retrieved May 4, 2016.

- "The Chinese Scientific Genius. Discoveries and inventions of an ancient civilization: Biological Pest Control" (PDF). The Courier. UNESCO: 24. October 1988. Retrieved 5 June 2016.

- Wilson, L. Ted; Pickett, Charles H.; Flaherty, Donald L.; Bates, Teresa A. "French prune trees: refuge for grape leafhopper parasite" (PDF). University of California Davis. Retrieved 7 June 2016.

- "Electronic Rodent Repellent Devices: A Review of Efficacy Test Protocols and Regulatory Actions". DigitalCommons@University of Nebraska - Lincoln. University of Nebraska - Lincoln.

Retrieved 8 December 2014.

- "Types of Pest Control Methods". brooklynpestcontrolservices.com. Brooklyn Pest Control Services. June 27, 2013. Retrieved 8 March 2014.

- "Cucurbitaceae--Fruits for Peons, Pilgrims, and Pharaohs". University of California at Los Angeles. Retrieved September 2, 2013.

- "Pacific Northwest Nursery IPM. Flowers, Sweets and a Nice Place to Stay: Courting Beneficials to Your Nursery". Oregon State University. Retrieved 11 February 2013.

Pesticides and Insecticides

The substances used for attracting, seducing and then destroying any pest is known as pesticides. Pesticides are commonly used for plant protection, while an insecticide is a substance used to kill insects. The aim of this chapter is to provide the readers an in-depth understanding on pesticides and insecticides and their importance in agriculture.

Pesticide

A crop-duster spraying pesticide on a field

Pesticides are substances meant for attracting, seducing, and then destroying any pest. They are a class of biocide. The most common use of pesticides is as plant protection products (also known as crop protection products), which in general protect plants from damaging influences such as weeds, fungi, or insects. This use of pesticides is so common that the term *pesticide* is often treated as synonymous with *plant protection product*, although it is in fact a broader term, as pesticides are also used for non-agricultural purposes. The term pesticide includes all of the following: herbicide, insecticide, insect growth regulator, nematicide, termiticide, molluscicide, piscicide, avicide, rodenticide, predacide, bactericide, insect repellent, animal repellent, antimicrobial, fungicide, disinfectant (antimicrobial), and sanitizer.

A Lite-Trac four-wheeled self-propelled crop sprayer spraying pesticide on a field

In general, a pesticide is a chemical or biological agent (such as a virus, bacterium, antimicrobial, or disinfectant) that deters, incapacitates, kills, or otherwise discourages pests. Target pests can include insects, plant pathogens, weeds, mollusks, birds, mammals, fish, nematodes (roundworms), and microbes that destroy property, cause nuisance, or spread disease, or are disease vectors. Although pesticides have benefits, some also have drawbacks, such as potential toxicity to humans and other species. According to the Stockholm Convention on Persistent Organic Pollutants, 9 of the 12 most dangerous and persistent organic chemicals are organochlorine pesticides.

Definition

Type of pesticide	Target pest group
Herbicides	Plant
Algicides or Algaecides	Algae
Avicides	Birds
Bactericides	Bacteria
Fungicides	Fungi and Oomycetes
Insecticides	Insects
Miticides or Acaricides	Mites
Molluscicides	Snails
Nematicides	Nematodes
Rodenticides	Rodents
Virucides	Viruses

The Food and Agriculture Organization (FAO) has defined *pesticide* as:

any substance or mixture of substances intended for preventing, destroying, or controlling any pest, including vectors of human or animal disease, unwanted species of plants or animals, causing harm during or otherwise interfering with the production, processing, storage, transport, or marketing of food, agricultural commodities, wood and wood products or animal feedstuffs, or substances that may be administered to animals for the control of insects, arachnids, or other pests in or on their bodies. The term includes substances intended for use as a plant growth regulator, defoliant, desiccant, or agent for thinning fruit or preventing the premature fall of fruit. Also used as substances applied to crops either before or after harvest to protect the commodity from deterioration during storage and transport.

Pesticides can be classified by target organism (e.g., herbicides, insecticides, fungicides, rodenticides, and pediculicides), chemical structure (e.g., organic, inorganic, synthetic, or biological (biopesticide), although the distinction can sometimes blur), and physical state (e.g. gaseous (fumigant)). Biopesticides include microbial pesticides and biochemical pesticides. Plant-derived pesticides, or "botanicals", have been developing quickly. These include the pyrethroids, rotenoids, nicotinoids, and a fourth group that includes strychnine and scilliroside.

Many pesticides can be grouped into chemical families. Prominent insecticide families include organochlorines, organophosphates, and carbamates. Organochlorine hydrocarbons (e.g., DDT) could be separated into dichlorodiphenylethanes, cyclodiene compounds, and other related compounds. They operate by disrupting the sodium/potassium balance of the nerve fiber, forcing the nerve to transmit continuously. Their toxicities vary greatly, but they have been phased out because of their persistence and potential to bioaccumulate. Organophosphate and carbamates largely replaced organochlorines. Both operate through inhibiting the enzyme acetylcholinesterase, allowing acetylcholine to transfer nerve impulses indefinitely and causing a variety of symptoms such as weakness or paralysis. Organophosphates are quite toxic to vertebrates, and have in some cases been replaced by less toxic carbamates. Thiocarbamate and dithiocarbamates are subclasses of carbamates. Prominent families of herbicides include phenoxy and benzoic acid herbicides (e.g. 2,4-D), triazines (e.g., atrazine), ureas (e.g., diuron), and Chloroacetanilides (e.g., alachlor). Phenoxy compounds tend to selectively kill broad-leaf weeds rather than grasses. The phenoxy and benzoic acid herbicides function similar to plant growth hormones, and grow cells without normal cell division, crushing the plant's nutrient transport system. Triazines interfere with photosynthesis. Many commonly used pesticides are not included in these families, including glyphosate.

Pesticides can be classified based upon their biological mechanism function or application method. Most pesticides work by poisoning pests. A systemic pesticide moves inside a plant following absorption by the plant. With insecticides and most fungicides, this movement is usually upward (through the xylem) and outward. Increased efficiency may be a result. Systemic insecticides, which poison pollen and nectar in the flowers, may kill bees and other needed pollinators.

In 2009, the development of a new class of fungicides called paldoxins was announced. These work by taking advantage of natural defense chemicals released by plants called phytoalexins, which fungi then detoxify using enzymes. The paldoxins inhibit the fungi's detoxification enzymes. They are believed to be safer and greener.

Uses

Pesticides are used to control organisms that are considered to be harmful. For example, they are used to kill mosquitoes that can transmit potentially deadly diseases like West Nile virus, yellow fever, and malaria. They can also kill bees, wasps or ants that can cause allergic reactions. Insecticides can protect animals from illnesses that can be caused by parasites such as fleas. Pesticides can prevent sickness in humans that could be caused by moldy food or diseased produce. Herbicides can be used to clear roadside weeds, trees and brush. They can also kill invasive weeds that may cause environmental damage. Herbicides are commonly applied in ponds and lakes to control algae and plants such as water grasses that can interfere with activities like swimming and fishing and cause the water to look or smell unpleasant. Uncontrolled pests such as termites and mold can damage structures such as houses. Pesticides are used in grocery stores and food storage facilities to manage rodents and insects that infest food such as grain. Each use of a pesticide carries some associated risk. Proper pesticide use decreases these associated risks to a level deemed acceptable by pesticide regulatory agencies such as the United States Environmental Protection Agency (EPA) and the Pest Management Regulatory Agency (PMRA) of Canada.

DDT, sprayed on the walls of houses, is an organochlorine that has been used to fight malaria since the 1950s. Recent policy statements by the World Health Organization have given stronger support to this approach. However, DDT and other organochlorine pesticides have been banned in most countries worldwide because of their persistence in the environment and human toxicity. DDT use is not always effective, as resistance to DDT was identified in Africa as early as 1955, and by 1972 nineteen species of mosquito worldwide were resistant to DDT.

Amount Used

In 2006 and 2007, the world used approximately 2.4 megatonnes (5.3×10^9 lb) of pesticides, with herbicides constituting the biggest part of the world pesticide use at 40%, followed by insecticides (17%) and fungicides (10%). In 2006 and 2007 the U.S. used approximately 0.5 megatonnes (1.1×10^9 lb) of pesticides, accounting for 22% of the world total, including 857 million pounds (389 kt) of conventional pesticides, which are used in the agricultural sector (80% of conventional pesticide use) as well as the industrial, commercial, governmental and home & garden sectors.Pesticides are also found in majority of U.S. households with 78 million out of the 105.5 million households indicating that they use some form of pesticide. As of 2007, there were more than

1,055 active ingredients registered as pesticides, which yield over 20,000 pesticide products that are marketed in the United States.

The US used some 1 kg (2.2 pounds) per hectare of arable land compared with: 4.7 kg in China, 1.3 kg in the UK, 0.1 kg in Cameroon, 5.9 kg in Japan and 2.5 kg in Italy. Insecticide use in the US has declined by more than half since 1980, (.6%/yr) mostly due to the near phase-out of organophosphates. In corn fields, the decline was even steeper, due to the switchover to transgenic Bt corn.

For the global market of crop protection products, market analysts forecast revenues of over 52 billion US$ in 2019.

Benefits

Pesticides can save farmers' money by preventing crop losses to insects and other pests; in the U.S., farmers get an estimated fourfold return on money they spend on pesticides. One study found that not using pesticides reduced crop yields by about 10%. Another study, conducted in 1999, found that a ban on pesticides in the United States may result in a rise of food prices, loss of jobs, and an increase in world hunger.

There are two levels of benefits for pesticide use, primary and secondary. Primary benefits are direct gains from the use of pesticides and secondary benefits are effects that are more long-term.

Primary Benefits

1. Controlling pests and plant disease vectors

    - Improved crop/livestock yields

    - Improved crop/livestock quality

    - Invasive species controlled

2. Controlling human/livestock disease vectors and nuisance organisms

    - Human lives saved and suffering reduced

    - Animal lives saved and suffering reduced

    - Diseases contained geographically

3. Controlling organisms that harm other human activities and structures

    - Drivers view unobstructed

    - Tree/brush/leaf hazards prevented

    - Wooden structures protected

Monetary

Every dollar ($1) that is spent on pesticides for crops yields four dollars ($4) in crops saved. This means based that, on the amount of money spent per year on pesticides, $10 billion, there is an additional $40 billion savings in crop that would be lost due to damage by insects and weeds. In general, farmers benefit from having an increase in crop yield and from being able to grow a variety of crops throughout the year. Consumers of agricultural products also benefit from being able to afford the vast quantities of produce available year-round. The general public also benefits from the use of pesticides for the control of insect-borne diseases and illnesses, such as malaria. The use of pesticides creates a large job market within the agrichemical sector.

Costs

On the cost side of pesticide use there can be costs to the environment, costs to human health, as well as costs of the development and research of new pesticides.

Health Effects

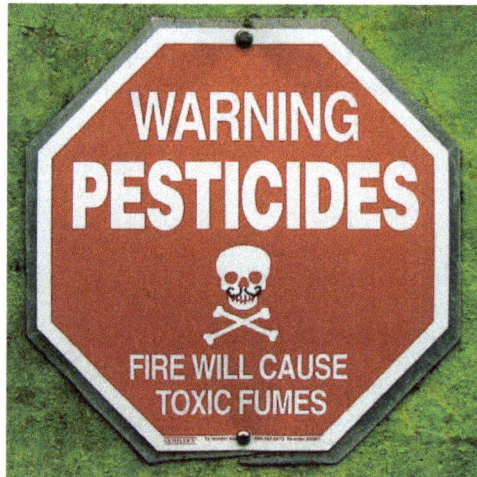

A sign warning about potential pesticide exposure.

Pesticides may cause acute and delayed health effects in people who are exposed. Pesticide exposure can cause a variety of adverse health effects, ranging from simple irritation of the skin and eyes to more severe effects such as affecting the nervous system, mimicking hormones causing reproductive problems, and also causing cancer. A 2007 systematic review found that "most studies on non-Hodgkin lymphoma and leukemia showed positive associations with pesticide exposure" and thus concluded that cosmetic use of pesticides should be decreased. There is substantial evidence of associations between organophosphate insecticide exposures and neurobehavioral alterations. Limited evidence also exists for other negative outcomes from pesticide exposure including neurological, birth defects, and fetal death.

The American Academy of Pediatrics recommends limiting exposure of children to pesticides and using safer alternatives:

The World Health Organization and the UN Environment Programme estimate that each year, 3 million workers in agriculture in the developing world experience severe poisoning from pesticides, about 18,000 of whom die. Owing to inadequate regulation and safety precautions, 99% of pesticide related deaths occur in developing countries that account for only 25% of pesticide usage. According to one study, as many as 25 million workers in developing countries may suffer mild pesticide poisoning yearly. There are several careers aside from agriculture that may also put individuals at risk of health effects from pesticide exposure including pet groomers, groundskeepers, and fumigators.

One study found pesticide self-poisoning the method of choice in one third of suicides worldwide, and recommended, among other things, more restrictions on the types of pesticides that are most harmful to humans.

A 2014 epidemiological review found associations between autism and exposure to certain pesticides, but noted that the available evidence was insufficient to conclude that the relationship was causal.

Environmental Effect

Pesticide use raises a number of environmental concerns. Over 98% of sprayed insecticides and 95% of herbicides reach a destination other than their target species, including non-target species, air, water and soil. Pesticide drift occurs when pesticides suspended in the air as particles are carried by wind to other areas, potentially contaminating them. Pesticides are one of the causes of water pollution, and some pesticides are persistent organic pollutants and contribute to soil contamination.

In addition, pesticide use reduces biodiversity, contributes to pollinator decline, destroys habitat (especially for birds), and threatens endangered species. Pests can develop a resistance to the pesticide (pesticide resistance), necessitating a new pesticide. Alternatively a greater dose of the pesticide can be used to counteract the resistance, although this will cause a worsening of the ambient pollution problem.

Since chlorinated hydrocarbon pesticides dissolve in fats and are not excreted, organisms tend to retain them almost indefinitely. Biological magnification is the process whereby these chlorinated hydrocarbons (pesticides) are more concentrated at each level of the food chain. Among marine animals, pesticide concentrations are higher in carnivorous fishes, and even more so in the fish-eating birds and mammals at the top of the ecological pyramid. Global distillation is the process whereby pesticides are transported from warmer to colder regions of the Earth, in particular the Poles and mountain tops. Pesticides that evaporate into the atmosphere at relatively high temperature can be carried considerable distances (thousands of kilometers) by the wind to an area of lower temperature, where they condense and are carried back to the ground in rain or snow.

In order to reduce negative impacts, it is desirable that pesticides be degradable or at least quickly deactivated in the environment. Such loss of activity or toxicity of pesticides is due to both innate chemical properties of the compounds and environmental processes or conditions. For example, the presence of halogens within a chemical structure often slows down degradation in an aerobic environment. Adsorption to soil may retard pesticide movement, but also may reduce bioavailability to microbial degraders.

Economics

Harm	Annual US cost
Public health	$1.1 billion
Pesticide resistance in pest	$1.5 billion
Crop losses caused by pesticides	$1.4 billion
Bird losses due to pesticides	$2.2 billion
Groundwater contamination	$2.0 billion
Other costs	$1.4 billion
Total costs	$9.6 billion

Human health and environmental cost from pesticides in the United States is estimated at $9.6 billion offset by about $40 billion in increased agricultural production:

Additional costs include the registration process and the cost of purchasing pesticides. The registration process can take several years to complete (there are 70 different types of field test) and can cost $50–70 million for a single pesticide. Annually the United States spends $10 billion on pesticides.

Alternatives

Alternatives to pesticides are available and include methods of cultivation, use of biological pest controls (such as pheromones and microbial pesticides), genetic engineering, and methods of interfering with insect breeding. Application of composted yard waste has also been used as a way of controlling pests. These methods are becoming increasingly popular and often are safer than traditional chemical pesticides. In addition, EPA is registering reduced-risk conventional pesticides in increasing numbers.

Cultivation practices include polyculture (growing multiple types of plants), crop rotation, planting crops in areas where the pests that damage them do not live, timing planting according to when pests will be least problematic, and use of trap crops that attract pests away from the real crop. In the U.S., farmers have had success controlling insects by spraying with hot water at a cost that is about the same as pesticide spraying.

Release of other organisms that fight the pest is another example of an alternative to pesticide use. These organisms can include natural predators or parasites of the pests. Biological pesticides based on entomopathogenic fungi, bacteria and viruses cause disease in the pest species can also be used.

Interfering with insects' reproduction can be accomplished by sterilizing males of the target species and releasing them, so that they mate with females but do not produce offspring. This technique was first used on the screwworm fly in 1958 and has since been used with the medfly, the tsetse fly, and the gypsy moth. However, this can be a costly, time consuming approach that only works on some types of insects.

Agroecology emphasize nutrient recycling, use of locally available and renewable resources, adaptation to local conditions, utilization of microenvironments, reliance on indigenous knowledge and yield maximization while maintaining soil productivity. Agroecology also emphasizes empowering people and local communities to contribute to development, and encouraging "multi-directional" communications rather than the conventional "top-down" method.

Push Pull Strategy

The term "push-pull" was established in 1987 as an approach for integrated pest management (IPM). This strategy uses a mixture of behavior-modifying stimuli to manipulate the distribution and abundance of insects. "Push" means the insects are repelled or deterred away from whatever resource that is being protected. "Pull" means that certain stimuli (semiochemical stimuli, pheromones, food additives, visual stimuli, genetically altered plants, etc.) are used to attract pests to trap crops where they will be killed. There are numerous different components involved in order to implement a Push-Pull Strategy in IPM.

Many case studies testing the effectiveness of the push-pull approach have been done across the world. The most successful push-pull strategy was developed in Africa for subsistence farming. Another successful case study was performed on the control of *Helicoverpa* in cotton crops in Australia. In Europe, the Middle East, and the United States, push-pull strategies were successfully used in the controlling of *Sitona lineatus* in bean fields.

Some advantages of using the push-pull method are less use of chemical or biological materials and better protection against insect habituation to this control method. Some disadvantages of the push-pull strategy is that if there is a lack of appropriate knowledge of behavioral and chemical ecology of the host-pest interactions then this method becomes unreliable. Furthermore, because the push-pull method is not a very popular method of IPM operational and registration costs are higher.

Effectiveness

Some evidence shows that alternatives to pesticides can be equally effective as the use of chemicals. For example, Sweden has halved its use of pesticides with hardly any reduction in crops. In Indonesia, farmers have reduced pesticide use on rice fields by 65% and experienced a 15% crop increase. A study of Maize fields in northern Florida found that the application of composted yard waste with high carbon to nitrogen ratio to agricultural fields was highly effective at reducing the population of plant-parasitic nematodes and increasing crop yield, with yield increases ranging from 10% to 212%; the observed effects were long-term, often not appearing until the third season of the study.

However, pesticide resistance is increasing. In the 1940s, U.S. farmers lost only 7% of their crops to pests. Since the 1980s, loss has increased to 13%, even though more pesticides are being used. Between 500 and 1,000 insect and weed species have developed pesticide resistance since 1945.

Types

Pesticides are often referred to according to the type of pest they control. Pesticides can also be considered as either biodegradable pesticides, which will be broken down by microbes and other living beings into harmless compounds, or persistent pesticides, which may take months or years before they are broken down: it was the persistence of DDT, for example, which led to its accumulation in the food chain and its killing of birds of prey at the top of the food chain. Another way to think about pesticides is to consider those that are chemical pesticides or are derived from a common source or production method.

Some examples of chemically-related pesticides are:

Organophosphate Pesticides

Organophosphates affect the nervous system by disrupting, acetylcholinesterase activity, the enzyme that regulates acetylcholine, a neurotransmitter. Most organophosphates are insecticides. They were developed during the early 19th century, but their effects on insects, which are similar to their effects on humans, were discovered in 1932. Some are very poisonous. However, they usually are not persistent in the environment.

Carbamate Pesticides

Carbamate pesticides affect the nervous system by disrupting an enzyme that regulates acetylcholine, a neurotransmitter. The enzyme effects are usually reversible. There are several subgroups within the carbamates.

Organochlorine Insecticides

They were commonly used in the past, but many have been removed from the market due to their health and environmental effects and their persistence (e.g., DDT, chlordane, and toxaphene).

Pyrethroid Pesticides

They were developed as a synthetic version of the naturally occurring pesticide pyrethrin, which is found in chrysanthemums. They have been modified to increase their stability in the environment. Some synthetic pyrethroids are toxic to the nervous system.

Sulfonylurea Herbicides

The following sulfonylureas have been commercialized for weed control: amidosulfuron, azimsulfuron, bensulfuron-methyl, chlorimuron-ethyl, ethoxysulfuron, flazasulfuron, flupyrsulfuron-methyl-sodium, halosulfuron-methyl, imazosulfuron, nicosulfuron, oxasulfuron, primisulfuron-methyl, pyrazosulfuron-ethyl, rimsulfuron, sulfometuron-methyl Sulfosulfuron, terbacil, bispyribac-sodium, cyclosulfamuron, and pyrithiobac-sodium. Nicosulfuron, triflusulfuron methyl, and chlorsulfuron are broad-spectrum herbicides that kill plants by inhibiting the enzyme acetolactate synthase. In the 1960s, more than 1 kg/ha (0.89 lb/acre) crop protection chemical was typically applied, while sulfonylureates allow as little as 1% as much material to achieve the same effect.

Biopesticides

Biopesticides are certain types of pesticides derived from such natural materials as animals, plants, bacteria, and certain minerals. For example, canola oil and baking soda have pesticidal applications and are considered biopesticides. Biopesticides fall into three major classes:

Microbial pesticides which consist of bacteria, entomopathogenic fungi or viruses (and sometimes includes the metabolites that bacteria or fungi produce). Entomopathogenic nematodes are also often classed as microbial pesticides, even though they are multi-cellular.

Biochemical pesticides or herbal pesticides are naturally occurring substances that control (or monitor in the case of pheromones) pests and microbial diseases.

Plant-incorporated protectants (PIPs) have genetic material from other species incorporated into their genetic material (*i.e.* GM crops). Their use is controversial, especially in many European countries.

Classified by Type of Pest

Pesticides that are related to the type of pests are:

Type	Action
Algicides	Control algae in lakes, canals, swimming pools, water tanks, and other sites
Antifouling agents	Kill or repel organisms that attach to underwater surfaces, such as boat bottoms
Antimicrobials	Kill microorganisms (such as bacteria and viruses)
Attractants	Attract pests (for example, to lure an insect or rodent to a trap). (However, food is not considered a pesticide when used as an attractant.)
Biopesticides	Biopesticides are certain types of pesticides derived from such natural materials as animals, plants, bacteria, and certain minerals
Biocides	Kill microorganisms
Disinfectants and sanitizers	Kill or inactivate disease-producing microorganisms on inanimate objects
Fungicides	Kill fungi (including blights, mildews, molds, and rusts)
Fumigants	Produce gas or vapor intended to destroy pests in buildings or soil
Herbicides	Kill weeds and other plants that grow where they are not wanted
Insecticides	Kill insects and other arthropods
Miticides	Kill mites that feed on plants and animals
Microbial pesticides	Microorganisms that kill, inhibit, or out compete pests, including insects or other microorganisms
Molluscicides	Kill snails and slugs
Nematicides	Kill nematodes (microscopic, worm-like organisms that feed on plant roots)
Ovicides	Kill eggs of insects and mites
Pheromones	Biochemicals used to disrupt the mating behavior of insects
Repellents	Repel pests, including insects (such as mosquitoes) and birds
Rodenticides	Control mice and other rodents

Further Types of Pesticides

The term pesticide also include these substances:

Defoliants : Cause leaves or other foliage to drop from a plant, usually to facilitate harvest. Desiccants : Promote drying of living tissues, such as unwanted plant tops. Insect growth regulators: Disrupt the molting, maturity from pupal stage to adult, or other life processes of insects.

Plant growth regulators : Substances (excluding fertilizers or other plant nutrients) that alter the expected growth, flowering, or reproduction rate of plants.

Regulation

International

In most countries, pesticides must be approved for sale and use by a government agency.

In Europe, recent EU legislation has been approved banning the use of highly toxic pesticides including those that are carcinogenic, mutagenic or toxic to reproduction, those that are endocrine-disrupting, and those that are persistent, bioaccumulative and toxic (PBT) or very persistent and very bioaccumulative (vPvB). Measures were approved to improve the general safety of pesticides across all EU member states.

Though pesticide regulations differ from country to country, pesticides, and products on which they were used are traded across international borders. To deal with inconsistencies in regulations among countries, delegates to a conference of the United Nations Food and Agriculture Organization adopted an International Code of Conduct on the Distribution and Use of Pesticides in 1985 to create voluntary standards of pesticide regulation for different countries. The Code was updated in 1998 and 2002. The FAO claims that the code has raised awareness about pesticide hazards and decreased the number of countries without restrictions on pesticide use.

Three other efforts to improve regulation of international pesticide trade are the United Nations London Guidelines for the Exchange of Information on Chemicals in International Trade and the United Nations Codex Alimentarius Commission. The former seeks to implement procedures for ensuring that prior informed consent exists between countries buying and selling pesticides, while the latter seeks to create uniform standards for maximum levels of pesticide residues among participating countries. Both initiatives operate on a voluntary basis.

Pesticides safety education and pesticide applicator regulation are designed to protect the public from pesticide misuse, but do not eliminate all misuse. Reducing the use of pesticides and choosing less toxic pesticides may reduce risks placed on society and the environment from pesticide use. Integrated pest management, the use of multiple approaches to control pests, is becoming widespread and has been used with success in countries such as Indonesia, China, Bangladesh, the U.S., Australia, and Mexico. IPM attempts to recognize the more widespread impacts of an action on an ecosystem, so that natural balances are not upset. New pesticides are being developed, including biological and botanical derivatives and alternatives that are thought to reduce health and environmental risks. In addition, applicators are being encouraged to consider alternative controls and adopt methods that reduce the use of chemical pesticides.

Pesticides can be created that are targeted to a specific pest's lifecycle, which can be environmentally more friendly. For example, potato cyst nematodes emerge from their protective cysts in response to a chemical excreted by potatoes; they feed on the potatoes and damage the crop. A similar chemical can be applied to fields early, before the potatoes are planted, causing the nematodes to emerge early and starve in the absence of potatoes.

United States

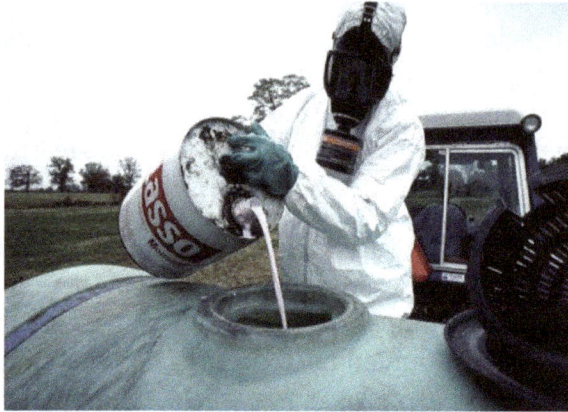

Preparation for an application of hazardous herbicide in USA.

In the United States, the Environmental Protection Agency (EPA) is responsible for regulating pesticides under the Federal Insecticide, Fungicide, and Rodenticide Act (FIFRA) and the Food Quality Protection Act (FQPA). Studies must be conducted to establish the conditions in which the material is safe to use and the effectiveness against the intended pest(s). The EPA regulates pesticides to ensure that these products do not pose adverse effects to humans or the environment. Pesticides produced before November 1984 continue to be reassessed in order to meet the current scientific and regulatory standards. All registered pesticides are reviewed every 15 years to ensure they meet the proper standards. During the registration process, a label is created. The label contains directions for proper use of the material in addition to safety restrictions. Based on acute toxicity, pesticides are assigned to a Toxicity Class.

Some pesticides are considered too hazardous for sale to the general public and are designated restricted use pesticides. Only certified applicators, who have passed an exam, may purchase or supervise the application of restricted use pesticides. Records of sales and use are required to be maintained and may be audited by government agencies charged with the enforcement of pesticide regulations. These records must be made available to employees and state or territorial environmental regulatory agencies.

The EPA regulates pesticides under two main acts, both of which amended by the Food Quality Protection Act of 1996. In addition to the EPA, the United States Department of Agriculture (USDA) and the United States Food and Drug Administration (FDA) set standards for the level of pesticide residue that is allowed on or in crops. The EPA looks at what the potential human health and environmental effects might be associated with the use of the pesticide.

In addition, the U.S. EPA uses the National Research Council's four-step process for human health risk assessment: (1) Hazard Identification, (2) Dose-Response Assessment, (3) Exposure Assessment, and (4) Risk Characterization.

Recently Kaua'i County (Hawai'i) passed Bill No. 2491 to add an article to Chapter 22 of the county's code relating to pesticides and GMOs. The bill strengthens protections of local communities in Kaua'i where many large pesticide companies test their products.

History

Since before 2000 BC, humans have utilized pesticides to protect their crops. The first known pesticide was elemental sulfur dusting used in ancient Sumer about 4,500 years ago in ancient Mesopotamia. The Rig Veda, which is about 4,000 years old, mentions the use of poisonous plants for pest control. By the 15th century, toxic chemicals such as arsenic, mercury, and lead were being applied to crops to kill pests. In the 17th century, nicotine sulfate was extracted from tobacco leaves for use as an insecticide. The 19th century saw the introduction of two more natural pesticides, pyrethrum, which is derived from chrysanthemums, and rotenone, which is derived from the roots of tropical vegetables. Until the 1950s, arsenic-based pesticides were dominant. Paul Müller discovered that DDT was a very effective insecticide. Organochlorines such as DDT were dominant, but they were replaced in the U.S. by organophosphates and carbamates by 1975. Since then, pyrethrin compounds have become the dominant insecticide. Herbicides became common in the 1960s, led by "triazine and other nitrogen-based compounds, carboxylic acids such as 2,4-dichlorophenoxyacetic acid, and glyphosate".

The first legislation providing federal authority for regulating pesticides was enacted in 1910; however, decades later during the 1940s manufacturers began to produce large amounts of synthetic pesticides and their use became widespread. Some sources consider the 1940s and 1950s to have been the start of the "pesticide era." Although the U.S. Environmental Protection Agency was established in 1970 and amendments to the pesticide law in 1972, pesticide use has increased 50-fold since 1950 and 2.3 million tonnes (2.5 million short tons) of industrial pesticides are now used each year. Seventy-five percent of all pesticides in the world are used in developed countries, but use in developing countries is increasing. A study of USA pesticide use trends through 1997 was published in 2003 by the National Science Foundation's Center for Integrated Pest Management.

In the 1960s, it was discovered that DDT was preventing many fish-eating birds from reproducing, which was a serious threat to biodiversity. Rachel Carson wrote the best-selling book *Silent Spring* about biological magnification. The agricultural use of DDT is now banned under the Stockholm Convention on Persistent Organic Pollutants, but it is still used in some developing nations to prevent malaria and other tropical diseases by spraying on interior walls to kill or repel mosquitoes.

Insecticide

An insecticide is a substance used to kill insects. They include ovicides and larvicides used against insect eggs and larvae, respectively. Insecticides are used in agriculture, medicine, industry and by consumers. Insecticides are claimed to be a major factor behind the increase in agricultural 20th century's productivity. Nearly all insecticides have the potential to significantly alter ecosystems; many are toxic to humans; some concentrate along the food chain.

Insecticides can be classified in two major groups: systemic insecticides, which have residual or long term activity; and contact insecticides, which have no residual activity.

Furthermore, one can distinguish three types of insecticide. 1. Natural insecticides, such as nicotine, pyrethrum and neem extracts, made by plants as defenses against insects. 2. Inorganic insecticides, which are metals. 3. Organic insecticides, which are organic chemical compounds, mostly working by contact.

The mode of action describes how the pesticide kills or inactivates a pest. It provides another way of classifying insecticides. Mode of action is important in understanding whether an insecticide will be toxic to unrelated species, such as fish, birds and mammals.

Insecticides are distinct from insect repellents, which do not kill.

Type of Activity

Systemic insecticides become incorporated and distributed systemically throughout the whole plant. When insects feed on the plant, they ingest the insecticide. Systemic insecticides produced by transgenic plants are called plant-incorporated protectants (PIPs). For instance, a gene that codes for a specific Bacillus thuringiensis biocidal protein was introduced into corn and other species. The plant manufactures the protein, which kills the insect when consumed. Systemic insecticides have activity pertaining to their residue which is called "residual activity" or long-term activity.

Contact insecticides are toxic to insects upon direct contact. These can be inorganic insecticides, which are metals and include arsenates, copper and fluorine compounds, which are less commonly used, and the commonly used sulfur. Contact insecticides can be organic insecticides, i.e. organic chemical compounds, synthetically produced, and comprising the largest numbers of pesticides used today. Or they can be natural compounds like pyrethrum, neem oil etc. Contact insecticides usually have no residual activity.

Efficacy can be related to the quality of pesticide application, with small droplets, such as aerosols often improving performance.

Biological Pesticides

Many organic compounds are produced by plants for the purpose of defending the host plant from predation. A trivial case is tree rosin, which is a natural insecticide. Specific, the production of oleoresin by conifer species is a component of the defense response against insect attack and fungal pathogen infection. Many fragrances, e.g. oil of wintergreen, are in fact antifeedants.

Four extracts of plants are in commercial use: pyrethrum, rotenone, neem oil, and various essential oils

Other Biological Approaches

Plant-Incorporated Protectants

Transgenic crops that act as insecticides began in 1996 with a genetically modified potato that produced the Cry protein, derived from the bacterium Bacillus thuringiensis, which is toxic to beetle larvae such as the Colorado potato beetle. The technique has been expanded to include the use of RNA interference RNAi that fatally silences crucial insect genes. RNAi likely evolved as a defense against viruses. Midgut cells in many larvae take up the molecules and help spread the signal. The technology can target only insects that have the silenced sequence, as was demonstrated when a particular RNAi affected only one of four fruit fly species. The technique is expected to replace many other insecticides, which are losing effectiveness due to the spread of pesticide resistance.

Enzymes

Many plants exude substances to repel insects. Premier examples are substances activated by the enzyme myrosinase. This enzyme converts glucosinolates to various compounds that are toxic to herbivorous insects. One product of this enzyme is allyl isothiocyanate, the pungent ingredient in horseradish sauces.

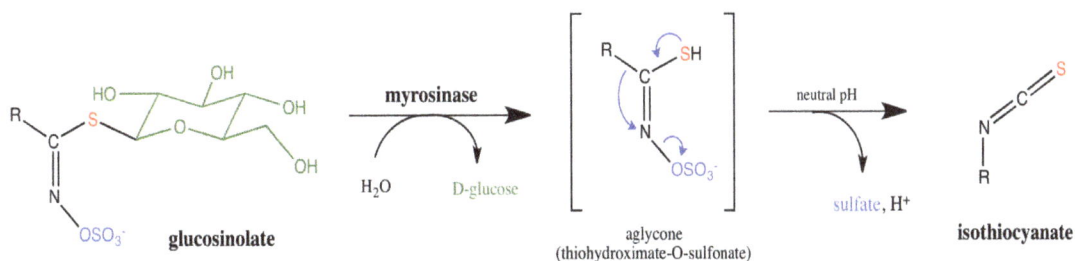

Biosynthesis of antifeedants by the action of myrosinase.

The myrosinase is released only upon crushing the flesh of horseradish. Since allyl isothiocyanate is harmful to the plant as well as the insect, it is stored in the harmless form of the glucosinolate, separate from the myrosinase enzyme.

Bacterial

Bacillus thuringiensis is a bacterial disease that affects Lepidopterans and some other insects. Toxins produced by strains of this bacterium are used as a larvicide against caterpillars, beetles, and mosquitoes. Toxins from *Saccharopolyspora spinosa* are isolated from fermentations and sold as Spinosad. Because these toxins have little effect on other organisms, they are considered more environmentally friendly than synthetic pesticides. The toxin from *B. thuringiensis* (Bt toxin) has been incorporated directly into plants through the use of genetic engineering. Other biological insecticides include products based on entomopathogenic fungi (e.g., *Beauveria bassiana*, *Metarhizium anisopliae*), nematodes (e.g., *Steinernema feltiae*) and viruses (e.g., *Cydia pomonella* granulovirus).

Synthetic Insecticide

A major emphasis of organic chemistry is the development of chemical tools to enhance agricultural productivity. Insecticides represent a major area of emphasis. Many of the major insecticides are inspired by biological analogues. Many others are completely alien to nature.

Organochlorides

The best known organochloride, DDT, was created by Swiss scientist Paul Müller. For this discovery, he was awarded the 1948 Nobel Prize for Physiology or Medicine. DDT was introduced in 1944. It functions by opening sodium channels in the insect's nerve cells. The contemporaneous rise of the chemical industry facilitated large-scale production of DDT and related chlorinated hydrocarbons.

Organophosphates and Carbamates

Organophosphates are another large class of contact insecticides. These also target the insect's nervous system. Organophosphates interfere with the enzymes acetylcholinesterase and other cholinesterases, disrupting nerve impulses and killing or disabling the insect. Organophosphate insecticides and chemical warfare nerve agents (such as sarin, tabun, soman, and VX) work in the same way. Organophosphates have a cumulative toxic effect to wildlife, so multiple exposures to the chemicals amplifies the toxicity. In the US, organophosphate use declined with the rise of substitutes.

Carbamate insecticides have similar mechanisms to organophosphates, but have a much shorter duration of action and are somewhat less toxic.

Pyrethroids

Pyrethroid pesticides mimic the insecticidal activity of the natural compound pyrethrum, the biopesticide found in pyrethrins. These compounds are nonpersistent sodium channel modulators and are less toxic than organophosphates and carbamates. Compounds in this group are often applied against household pests.

Neonicotinoids

Neonicotinoids are synthetic analogues of the natural insecticide nicotine (with much lower acute mammalian toxicity and greater field persistence). These chemicals are acetylcholine receptor agonists. They are broad-spectrum systemic insecticides, with rapid action (minutes-hours). They are applied as sprays, drenches, seed and soil treatments. Treated insects exhibit leg tremors, rapid wing motion, stylet withdrawal (aphids), disoriented movement, paralysis and death. Imidacloprid may be the most common. It has recently come under scrutiny for allegedly pernicious effects on honeybees and its potential to increase the susceptibility of rice to planthopper attacks.

Ryanoids

Ryanoids are synthetic analogues with the same mode of action as ryanodine, a naturally occurring insecticide extracted from *Ryania speciosa* (Flacourtiaceae). They bind to calcium channels in cardiac and skeletal muscle, blocking nerve transmission. Only one such insecticide is currently registered, Rynaxypyr, generic name chlorantraniliprole.

Insect Growth Regulators

Insect growth regulator (IGR) is a term coined to include insect hormone mimics and an earlier class of chemicals, the benzoylphenyl ureas, which inhibit chitin(exoskeleton) biosynthesis in insects. Diflubenzuron is a member of the latter class, used primarily to control caterpillars that are pests. The most successful insecticides in this class are the juvenoids (juvenile hormone analogues). Of these, methoprene is most widely used. It has no observable acute toxicity in rats and is approved by World Health Organization (WHO) for use in drinking water cisterns to combat malaria. Most of its uses are to combat insects where the adult is the pest, including mosquitoes, several fly species, and fleas. Two very similar products, hydroprene and kinoprene, are used for controlling species such as cockroaches and white flies. Methoprene was registered with the EPA in 1975. Virtually no reports of resistance have been filed. A more recent type of IGR is the ecdysone agonist tebufenozide (MIMIC), which is used in forestry and other applications for control of caterpillars, which are far more sensitive to its hormonal effects than other insect orders.

Environmental Effects

Effects on Nontarget Species

Some insecticides kill or harm other creatures in addition to those they are intended to kill. For example, birds may be poisoned when they eat food that was recently sprayed with insecticides or when they mistake an insecticide granule on the ground for food and eat it.

Sprayed insecticide may drift from the area to which it is applied and into wildlife areas, especially when it is sprayed aerially.

DDT

The development of DDT was motivated by desire to replace more dangerous or less effective alternatives. DDT was introduced to replace lead and arsenic-based compounds, which were in widespread use in the early 1940s.

DDT was brought to public attention by Rachel Carson's book *Silent Spring*. One side-effect of DDT is to reduce the thickness of shells on the eggs of predatory birds. The shells sometimes become too thin to be viable, reducing bird populations. This occurs with DDT and related compounds due to the process of bioaccumulation, wherein the chemical, due to its stability and fat solubility, accumulates in organisms' fatty tissues. Also, DDT may biomagnify, which causes progressively higher concentrations in the body fat of animals farther up the food chain. The near-world-wide ban on agricultural use of DDT and related chemicals has allowed some of these birds, such as the peregrine falcon, to recover in recent years. A number of organochlorine pesticides have been banned from most uses worldwide. Globally they are controlled via the Stockholm Convention on persistent organic pollutants. These include: aldrin, chlordane, DDT, dieldrin, endrin, heptachlor, mirex and toxaphene.

Pollinator decline

Insecticides can kill bees and may be a cause of pollinator decline, the loss of bees that pollinate plants, and colony collapse disorder (CCD), in which worker bees from a beehive or Western honey bee colony abruptly disappear. Loss of pollinators means a reduction in crop yields. Sublethal doses of insecticides (i.e. imidacloprid and other neonicotinoids) affect bee foraging behavior. However, research into the causes of CCD was inconclusive as of June 2007.

Agrochemical

Agrochemical or agrichemical, a contraction of *agricultural chemical*, is a generic term for the various chemical products used in agriculture. In most cases, *agrichemical* refers to the broad range of pesticides, including insecticides, herbicides, fungicides and nematicides. It may also include synthetic fertilizers, hormones and other chemical growth agents, and concentrated stores of raw animal manure.

The Passaic Agricultural Chemical Works in Newark, New Jersey, 1876

Ecology

Many agrichemicals are toxic, and agrichemicals in bulk storage may pose significant environmental and/or health risks, particularly in the event of accidental spills. In many countries, use of agrichemicals is highly regulated. Government-issued permits for purchase and use of approved agrichemicals may be required. Significant penalties can result from misuse, including improper storage resulting in spillage. On farms, proper storage facilities and labeling, emergency clean-up equipment and procedures, and safety equipment and procedures for handling, application and disposal are often subject to mandatory standards and regulations. Usually, the regulations are carried out through the registration process.

For instance, bovine somatotropin, though widely used in the U.S.A., is not approved in Canada and some other jurisdictions as there are concerns for the health of cows using it.

Companies

Syngenta was the worldwide leader in agrochemical sales in 2013 at ~$10.9 billion, followed by Bayer CropScience, BASF, Dow Agrosciences, Monsanto, and then DuPont with ~$3.6 billion.

Nematicide

A nematicide is a type of chemical pesticide used to kill plant-parasitic nematodes. Nematicides have tended to be broad-spectrum toxicants possessing high volatility or other properties promoting migration through the soil. Aldicarb (Temik), a carbamate insecticide marketed by Bayer CropScience, is an example of a commonly used commercial nematicide. It is important in potato production, where it has been used for control of soil-borne nematodes. Aldicarb is a cholinesterase inhibitor, which prevents the breakdown of acetylcholine in the synapse. In case of severe poisoning, the victim dies of respiratory failure. It is no longer authorised for use in the EU and, in August, 2010, Bayer CropScience announced that it plans to discontinue aldicarb by 2014. Human health safety and environmental concerns have resulted in the widespread deregistration of several other agronomically important nematicides. Prior to 1985, the persistent halocarbon DBCP was a widely used nematicide and soil fumigant. However, it was banned from use after being linked to sterility among male workers; the Dow Chemical company was subsequently found liable for more than $600 million in damages.

Several natural nematicides are known. An environmentally benign garlic-derived polysulfide product is approved for use in the European Union (under Annex 1 of 91/414) and the UK as a nematicide. Another common natural nematicide is obtained from neem cake, the residue obtained after cold-pressing the fruit and kernels of the neem tree. Known by several names in the world, the tree was first cultivated in India in ancient times and is now widely distributed throughout the world. The root exudate of marigold (*Tagetes*) is also found to have nematicidal action. Nematophagous fungi, a type of carnivorous fungi, can be useful in controlling nematodes, *Paecilomyces* being one example.

Besides chemicals, soil steaming can be used in order to kill nematodes. Superheated steam is induced into the soil, which causes almost all organic material to deteriorate.

Fungicide

Fungicides are biocidal chemical compounds or biological organisms used to kill fungi or fungal spores. A fungistatic inhibits their growth. Fungi can cause serious damage in agriculture, resulting in critical losses of yield, quality, and profit. Fungicides are used both in agriculture and to fight fungal infections in animals. Chemicals used to control oomycetes, which are not fungi, are also referred to as fungicides, as oomycetes use the same mechanisms as fungi to infect plants.

Fungicides can either be contact, translaminar or systemic. Contact fungicides are not taken up into the plant tissue and protect only the plant where the spray is deposited. Translaminar fungicides redistribute the fungicide from the upper, sprayed leaf surface to the lower, unsprayed surface. Systemic fungicides are taken up and redistributed through the xylem vessels. Few fungicides move to all parts of a plant. Some are locally systemic, and some move upwardly.

Most fungicides that can be bought retail are sold in a liquid form. A very common active ingredient is sulfur, present at 0.08% in weaker concentrates, and as high as 0.5% for more potent fungicides. Fungicides in powdered form are usually around 90% sulfur and are very toxic. Other active ingredients in fungicides include neem oil, rosemary oil, jojoba oil, the bacterium *Bacillus subtilis*, and the beneficial fungus *Ulocladium oudemansii*.

Fungicide residues have been found on food for human consumption, mostly from post-harvest treatments. Some fungicides are dangerous to human health, such as vinclozolin, which has now been removed from use. Ziram is also a fungicide that is thought to be toxic to humans if exposed to chronically. A number of fungicides are also used in human health care.

Natural Fungicides

Plants and other organisms have chemical defenses that give them an advantage against microorganisms such as fungi. Some of these compounds can be used as fungicides:

- Tea tree oil
- Cinnamaldehyde
- Citronella oil
- Jojoba oil
- Nimbin
- Oregano oil
- Rosemary oil
- Monocerin
- Milk

Whole live or dead organisms that are efficient at killing or inhibiting fungi can sometimes be used as fungicides:

- *Bacillus subtilis*
- *Ulocladium oudemansii*
- Kelp (powdered dried kelp is fed to cattle to help prevent fungal infection)
- *Ampelomyces quisqualis*

Resistance

Pathogens respond to the use of fungicides by evolving resistance. In the field several mechanisms of resistance have been identified. The evolution of fungicide resistance can be gradual or sudden. In qualitative or discrete resistance, a mutation (normally to a single gene) produces a race of a fungus with a high degree of resistance. Such resistant varieties also tend to show stability, persisting after the fungicide has been removed from the market. For example, sugar beet leaf blotch remains resistant to azoles years after they were no longer used for control of the disease. This is because such mutations often have a high selection pressure when the fungicide is used, but there is low selection pressure to remove them in the absence of the fungicide.

In instances where resistance occurs more gradually, a shift in sensitivity in the pathogen to the fungicide can be seen. Such resistance is polygenic – an accumulation of many mutations in different genes, each having a small additive effect. This type of resistance is known as quantitative or continuous resistance. In this kind of resistance, the pathogen population will revert to a sensitive state if the fungicide is no longer applied.

Little is known about how variations in fungicide treatment affect the selection pressure to evolve resistance to that fungicide. Evidence shows that the doses that provide the most control of the disease also provide the largest selection pressure to acquire resistance, and that lower doses decrease the selection pressure.

In some cases when a pathogen evolves resistance to one fungicide, it automatically obtains resistance to others – a phenomenon known as cross resistance. These additional fungicides are normally of the same chemical family or have the same mode of action, or can be detoxified by the same mechanism. Sometimes negative cross resistance occurs, where resistance to one chemical class of fungicides leads to an increase in sensitivity to a different chemical class of fungicides. This has been seen with carbendazim and diethofencarb.

There are also recorded incidences of the evolution of multiple drug resistance by pathogens – resistance to two chemically different fungicides by separate mutation events. For example, *Botrytis cinerea* is resistant to both azoles and dicarboximide fungicides.

There are several routes by which pathogens can evolve fungicide resistance. The most common mechanism appears to be alteration of the target site, in particular as a defence against single site of action fungicides. For example, Black Sigatoka, an economically important pathogen of banana, is resistant to the QoI fungicides, due to a single nucleotide change resulting in the replacement of one amino acid (glycine) by another (alanine) in the target protein of the QoI fungicides, cytochrome b. It is presumed that this disrupts the binding of the fungicide to the protein, rendering the fungicide ineffective. Upregulation of target genes can also render the fungicide ineffective. This is seen in DMI-resistant strains of *Venturia inaequalis*.

Resistance to fungicides can also be developed by efficient efflux of the fungicide out of the cell. *Septoria tritici* has developed multiple drug resistance using this mechanism. The pathogen had 5 ABC-type transporters with overlapping substrate specificities that together work to pump toxic chemicals out of the cell.

In addition to the mechanisms outlined above, fungi may also develop metabolic pathways that circumvent the target protein, or acquire enzymes that enable metabolism of the fungicide to a harmless substance.

Fungicide Resistance Management

The fungicide resistance action committee (FRAC) has several recommended practices to try to avoid the development of fungicide resistance, especially in at-risk fungicides including *Strobilurins* such as azoxystrobin.

Products should not be used in isolation, but rather as mixture, or alternate sprays, with another fungicide with a different mechanism of action. The likelihood of the pathogen's developing resistance is greatly decreased by the fact that any resistant isolates to one fungicide will be killed by the other; in other words, two mutations would be required rather than just one. The effectiveness of this technique can be demonstrated by Metalaxyl, a phenylamide fungicide. When used as the sole product in Ireland to control potato blight (*Phytophthora infestans*), resistance developed within one growing season. However, in countries like the UK where it was marketed only as a mixture, resistance problems developed more slowly.

Fungicides should be applied only when absolutely necessary, especially if they are in an at-risk group. Lowering the amount of fungicide in the environment lowers the selection pressure for resistance to develop.

Manufacturers' doses should always be followed. These doses are normally designed to give the right balance between controlling the disease and limiting the risk of resistance development. Higher doses increase the selection pressure for single-site mutations that confer resistance, as all strains but those that carry the mutation will be eliminated, and thus the resistant strain will propagate. Lower doses greatly increase the risk of polygenic resistance, as strains that are slightly less sensitive to the fungicide may survive.

It is also recommended that where possible fungicides are used only in a protective manner, rather than to try to cure already-infected crops. Far fewer fungicides have curative/eradicative ability than protectant. Thus, fungicide preparations advertised as having curative action may have only one active chemical; a single fungicide acting in isolation increases the risk of fungicide resistance.

It is better to use an integrative pest management approach to disease control rather than relying on fungicides alone. This involves the use of resistant varieties and hygienic practices, such as the removal of potato discard piles and stubble on which the pathogen can overwinter, greatly reducing the titre of the pathogen and thus the risk of fungicide resistance development.

Herbicide

Herbicide(s), also commonly known as weedkillers, are chemical substances used to control unwanted plants. Selective herbicides control specific weed species, while leaving the desired crop relatively unharmed, while non-selective herbicides (sometimes called "total weedkillers" in commercial products) can be used to clear waste ground, industrial and construction sites, railways and railway embankments as they kill all plant material with which they come into contact. Apart from selective/non-selective, other important distinctions include *persistence* (also known as *residual action*: how long the product stays in place and remains active), *means of uptake* (whether it is absorbed by above-ground foliage only, through the roots, or by other means), and *mechanism of action* (how it works). Historically, products such as common salt and other metal salts were used as herbicides, however these have gradually fallen out of favor and in some countries a number of these are banned due to their persistence in soil, and toxicity and groundwater contamination concerns. Herbicides have also been used in warfare and conflict.

Modern herbicides are often synthetic mimics of natural plant hormones which interfere with growth of the target plants. The term organic herbicide has come to mean herbicides intended for organic farming; these are often less efficient and more costly than synthetic herbicides and are based on natural materials. Some plants also produce their own natural herbicides, such as the genus *Juglans* (walnuts), or the tree of heaven; such action of natural herbicides, and other related chemical interactions, is called allelopathy. Due to herbicide resistance - a major concern in agriculture - a number of products also combine herbicides with different means of action.

In the US in 2007, about 83% of all herbicide usage, determined by weight applied, was in agriculture. In 2007, world pesticide expenditures totaled about $39.4 billion; herbicides were about 40% of those sales and constituted the biggest portion, followed by insecticides, fungicides, and other types. Smaller quantities are used in forestry, pasture systems, and management of areas set aside as wildlife habitat.

History

Prior to the widespread use of chemical herbicides, cultural controls, such as altering soil pH, salinity, or fertility levels, were used to control weeds. Mechanical control (including tillage) was also (and still is) used to control weeds.

First Herbicides

Although research into chemical herbicides began in the early 20th century, the first major breakthrough was the result of research conducted in both the UK and the US during the Second World War into the potential use of agents as biological weapons. The first modern herbicide, 2,4-D, was first discovered and synthesized by W. G. Templeman at Imperial Chemical Industries. In 1940, he showed that "Growth substances applied appropriately would kill certain broad-leaved weeds in cereals without harming the crops." By 1941, his team succeeded in synthesizing the chemical. In the same year, Pokorny in the US achieved this as well.

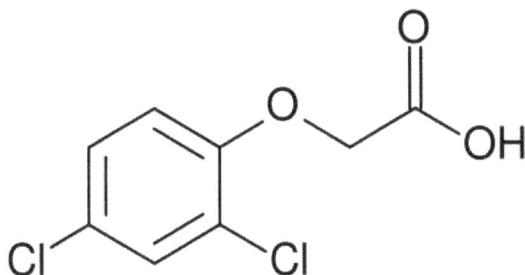

2,4-D, the first chemical herbicide, was discovered during the Second World War.

Independently, a team under Juda Hirsch Quastel, working at the Rothamsted Experimental Station made the same discovery. Quastel was tasked by the Agricultural Research Council (ARC) to discover methods for improving crop yield. By analyzing soil as a dynamic system, rather than an inert substance, he was able to apply techniques such as perfusion. Quastel was able to quantify the influence of various plant hormones, inhibitors and other chemicals on the activity of microorganisms in the soil and assess their direct impact on plant growth. While the full work of the unit remained secret, certain discoveries were developed for commercial use after the war, including the 2,4-D compound.

When it was commercially released in 1946, it triggered a worldwide revolution in agricultural output and became the first successful selective herbicide. It allowed for greatly enhanced weed control in wheat, maize (corn), rice, and similar cereal grass crops, because it kills dicots (broadleaf plants), but not most monocots (grasses). The low cost of 2,4-D has led to continued usage today, and it remains one of the most commonly used herbicides in the world. Like other acid herbicides, current formulations use either an amine salt (often trimethylamine) or one of many esters of the parent compound. These are easier to handle than the acid.

Further Discoveries

The triazine family of herbicides, which includes atrazine, were introduced in the 1950s; they have the current distinction of being the herbicide family of greatest concern regarding groundwater contamination. Atrazine does not break down readily (within a few weeks) after being applied to soils of above neutral pH. Under alkaline soil conditions, atrazine may be carried into the soil profile as far as the water table by soil water following rainfall causing the aforementioned contamination. Atrazine is thus said to have "carryover", a generally undesirable property for herbicides.

Glyphosate (Roundup) was introduced in 1974 for nonselective weed control. Following the development of glyphosate-resistant crop plants, it is now used very extensively for selective weed control in growing crops. The pairing of the herbicide with the resistant seed contributed to the consolidation of the seed and chemistry industry in the late 1990s.

Many modern chemical herbicides used in agriculture and gardening are specifically formulated to decompose within a short period after application. This is desirable, as it allows crops and plants to be planted afterwards, which could otherwise be affected by the herbicide. However, herbicides with low residual activity (i.e., that decompose quickly) often do not provide season-long weed control and do not ensure that weed roots are killed beneath construction and paving (and cannot emerge destructively in years to come), therefore there remains a role for weedkiller with high levels of persistence in the soil.

Terminology

Herbicides are classified/grouped in various ways e.g. according to the activity, timing of application, method of application, mechanism of action, chemical family. This gives rise to a considerable level of terminology related to herbicides and their use.

Intended Outcome

- Control is the destruction of unwanted weeds, or the damage of them to the point where they are no longer competitive with the crop.

- Suppression is incomplete control still providing some economic benefit, such as reduced competition with the crop.

- Crop safety, for selective herbicides, is the relative absence of damage or stress to the crop. Most selective herbicides cause some visible stress to crop plants.

- Defoliant, similar to herbicides, but designed to remove foliage (leaves) rather than kill the plant.

Selectivity (All Plants or Specific Plants)

- Selective herbicides: They control or suppress certain plants without affecting the growth of other plants species. Selectivity may be due to translocation, differential absorption, physical (morphological) or physiological differences between plant species. 2,4-D, mecoprop, dicamba control many broadleaf weeds but remain ineffective against turfgrasses.

- Non-selective herbicides: These herbicides are not specific in acting against certain plant species and control all plant material with which they come into contact. They are used to clear industrial sites, waste ground, railways and railway embankments. Paraquat, glufosinate, glyphosate are non-selective herbicides.

Timing of Application

- Preplant: Preplant herbicides are nonselective herbicides applied to soil before planting. Some preplant herbicides may be mechanically incorporated into the soil. The objective for incorporation is to prevent dissipation through photodecomposition and/or volatility. The herbicides kill weeds as they grow through the herbicide treated zone. Volatile herbicides have to be incorporated into the soil before planting the pasture. Agricultural crops grown in soil treated with a preplant herbicide include tomatoes, corn, soybeans and strawberries. Soil fumigants like metam-sodium and dazomet are in use as preplant herbicides.

- Preemergence: Preemergence herbicides are applied before the weed seedlings emerge through the soil surface. Herbicides do not prevent weeds from germinating but they kill weeds as they grow through the herbicide treated zone by affecting the cell division in the emerging seedling. Dithopyr and pendimethalin are preemergence herbicides. Weeds that have already emerged before application or activation are not affected by pre-herbicides as their primary growing point escapes the treatment.

- Postemergence: These herbicides are applied after weed seedlings have emerged through the soil surface. They can be foliar or root absorbed, selective or non-selective, contact or systemic. Application of these herbicides is avoided during rain because the problem of being washed off to the soil makes it ineffective. 2,4-D is a selective, systemic, foliar absorbed postemergence herbicide.

Method of Application

- Soil applied: Herbicides applied to the soil are usually taken up by the root or shoot of the emerging seedlings and are used as preplant or preemergence treatment. Several factors influence the effectiveness of soil-applied herbicides. Weeds absorb herbicides by both passive and active mechanism. Herbicide adsorption to soil colloids or organic matter often reduces its amount available for weed absorption. Positioning of herbicide in correct layer of soil is very important, which can be achieved mechanically and by rainfall. Herbicides on the soil surface are subjected to several processes that reduce their availability. Volatility and photolysis are two common processes that reduce the availability of herbicides. Many soil applied herbicides are absorbed through plant shoots while they are still underground leading to their death or injury. EPTC and trifluralin are soil applied herbicides.

- Foliar applied: These are applied to portion of the plant above the ground and are absorbed by exposed tissues. These are generally postemergence herbicides and can either be translocated (systemic) throughout the plant or remain at specific site (contact). External barriers of plants like cuticle, waxes, cell wall etc. affect herbicide absorption and action. Glyphosate, 2,4-D and dicamba are foliar applied herbicide.

Persistence

- Residual activity: A herbicide is described as having low residual activity if it is neutralized within a short time of application (within a few weeks or months) - typically this is due to rainfall, or by reactions in the soil. A herbicide described as having high residual activity will remains potent for a long term in the soil. For some compounds, the residual activity can leave the ground almost permanently barren.

Mechanism of Action

Herbicides are often classified according to their site of action, because as a general rule, herbicides within the same site of action class will produce similar symptoms on susceptible plants. Classification based on site of action of herbicide is comparatively better as herbicide resistance management can be handled more properly and effectively. Classification by mechanism of action (MOA) indicates the first enzyme, protein, or biochemical step affected in the plant following application.

List of Mechanisms Found in Modern Herbicides

- ACCase inhibitors compounds kill grasses. Acetyl coenzyme A carboxylase (AC-Case) is part of the first step of lipid synthesis. Thus, ACCase inhibitors affect cell membrane production in the meristems of the grass plant. The ACCases of grasses are sensitive to these herbicides, whereas the ACCases of dicot plants are not.

- ALS inhibitors: the acetolactate synthase (ALS) enzyme (also known as acetohydroxyacid synthase, or AHAS) is the first step in the synthesis of the branched-chain amino acids (valine, leucine, and isoleucine). These herbicides slowly starve affected plants of these amino acids, which eventually leads to inhibition of DNA synthesis. They affect grasses and dicots alike. The ALS inhibitor family includes various sulfonylureas (such as Flazasulfuron and Metsulfuron-methyl), imidazolinones, triazolopyrimidines, pyrimidinyl oxybenzoates, and sulfonylamino carbonyl triazolinones. The ALS biological pathway exists only in plants and not animals, thus making the ALS-inhibitors among the safest herbicides.

- EPSPS inhibitors: The enolpyruvylshikimate 3-phosphate synthase enzyme EPSPS is used in the synthesis of the amino acids tryptophan, phenylalanine and tyrosine. They affect grasses and dicots alike. Glyphosate (Roundup) is a systemic EPSPS inhibitor inactivated by soil contact.

- Synthetic auxins inaugurated the era of organic herbicides. They were discovered in the 1940s after a long study of the plant growth regulator auxin. Synthetic auxins mimic this plant hormone. They have several points of action on the cell membrane, and are effective in the control of dicot plants. 2,4-D is a synthetic auxin herbicide.

- Photosystem II inhibitors reduce electron flow from water to NADPH2+ at the photochemical step in photosynthesis. They bind to the Qb site on the D1 protein, and prevent quinone from binding to this site. Therefore, this group of compounds causes electrons to accumulate on chlorophyll molecules. As a consequence, oxidation reactions in excess of those normally tolerated by the cell occur, and the plant dies. The triazine herbicides (including atrazine) and urea derivatives (diuron) are photosystem II inhibitors.

- Photosystem I inhibitors steal electrons from the normal pathway through FeS to Fdx to NADP leading to direct discharge of electrons on oxygen. As a result, reactive oxygen species are produced and oxidation reactions in excess of those normally tolerated by the cell occur, leading to plant death. Bipyridinium herbicides (such as diquat and paraquat) inhibit the Fe-S – Fdx step of that chain, while diphenyl ether herbicides (such as nitrofen, nitrofluorfen, and acifluorfen) inhibit the Fdx – NADP step.

- HPPD inhibitors inhibit 4-Hydroxyphenylpyruvate dioxygenase, which are involved in tyrosine breakdown. Tyrosine breakdown products are used by plants to make carotenoids, which protect chlorophyll in plants from being destroyed by sunlight. If this happens, the plants turn white due to complete loss of chlorophyll, and the plants die. Mesotrione and sulcotrione are herbicides in this class; a drug, nitisinone, was discovered in the course of developing this class of herbicides.

Herbicide Group (Labeling)

One of the most important methods for preventing, delaying, or managing resistance is to reduce the reliance on a single herbicide mode of action. To do this, farmers must know the mode of action for the herbicides they intend to use, but the relatively complex nature of plant biochemistry makes this difficult to determine. Attempts were made to simplify the understanding of herbicide mode of action by developing a classification system that grouped herbicides by mode of action. Eventually the Herbicide Resistance Action Committee (HRAC) and the Weed Science Society of America (WSSA) developed a classification system. The WSSA and HRAC systems differ in the group designation. Groups in the WSSA and the HRAC systems are designated by numbers and letters, respectively. The goal for adding the "Group" classification and mode of action to the herbicide product label is to provide a simple and practical approach to deliver the information to users. This information will make it easier to develop educational material that is consistent and effective. It should increase user's awareness of herbicide mode of action and provide more accurate recommendations for resistance management. Another goal is to make it easier for users to keep records on which herbicide mode of actions are being used on a particular field from year to year.

Chemical Family

Detailed investigations on chemical structure of the active ingredients of the registered herbicides showed that some moieties (moiety is a part of a molecule that may include either whole functional groups or parts of functional groups as substructures; a functional group has similar chemical properties whenever it occurs in different compounds) have the same mechanisms of action. According to Forouzesh *et al.* 2015, these moieties have been assigned to the names of chemical families and active ingredients are then classified within the chemical families accordingly. Knowing about herbicide chemical family grouping could serve as a short-term strategy for managing resistance to site of action.

Use and Application

Herbicides being sprayed from the spray arms of a tractor in North Dakota.

Most herbicides are applied as water-based sprays using ground equipment. Ground equipment varies in design, but large areas can be sprayed using self-propelled sprayers equipped with long booms, of 60 to 120 feet (18 to 37 m) with spray nozzles spaced every 20–30 inches (510–760 mm) apart. Towed, handheld, and even horse-drawn sprayers are also used. On large areas, herbicides may also at times be applied aerially using helicopters or airplanes, or through irrigation systems (known as chemigation).

A further method of herbicide application developed around 2010, involves ridding the soil of its active weed seed bank rather than just killing the weed. This can successfully treat annual plants but not perennials. Researchers at the Agricultural Research Service found that the application of herbicides to fields late in the weeds' growing season greatly reduces their seed production, and therefore fewer weeds will return the following season. Because most weeds are annuals, their seeds will only survive in soil for a year or two, so this method will be able to destroy such weeds after a few years of herbicide application.

Weed-wiping may also be used, where a wick wetted with herbicide is suspended from a boom and dragged or rolled across the tops of the taller weed plants. This allows treatment of taller grassland weeds by direct contact without affecting related but desirable shorter plants in the grassland sward beneath. The method has the benefit of avoiding spray drift. In Wales, a scheme offering free weed-wiper hire was launched in 2015 in an effort to reduce the levels of MCPA in water courses.

Misuse and Misapplication

Herbicide volatilisation or spray drift may result in herbicide affecting neighboring fields or plants, particularly in windy conditions. Sometimes, the wrong field or plants may be sprayed due to error.

Health and Environmental Effects

Herbicides have widely variable toxicity in addition to acute toxicity from occupational exposure levels.

Some herbicides cause a range of health effects ranging from skin rashes to death. The pathway of attack can arise from intentional or unintentional direct consumption, improper application resulting in the herbicide coming into direct contact with people or wildlife, inhalation of aerial sprays, or food consumption prior to the labeled preharvest interval. Under some conditions, certain herbicides can be transported via leaching or surface runoff to contaminate groundwater or distant surface water sources. Generally, the conditions that promote herbicide transport include intense storm events (particularly shortly after application) and soils with limited capacity to adsorb or retain the herbicides. Herbicide properties that increase likelihood of transport include persistence (resistance to degradation) and high water solubility.

Phenoxy herbicides are often contaminated with dioxins such as TCDD; research has suggested such contamination results in a small rise in cancer risk after occupational exposure to these herbicides. Triazine exposure has been implicated in a likely relationship to increased risk of breast cancer, although a causal relationship remains unclear.

Herbicide manufacturers have at times made false or misleading claims about the safety of their products. Chemical manufacturer Monsanto Company agreed to change its advertising after pressure from New York attorney general Dennis Vacco; Vacco complained about misleading claims that its spray-on glyphosate-based herbicides, including Roundup, were safer than table salt and "practically non-toxic" to mammals, birds, and fish (though proof that this was ever said is hard to find). Roundup is toxic and has resulted in death after being ingested in quantities ranging from 85 to 200 ml, although it has also been ingested in quantities as large as 500 ml with only mild or moderate symptoms. The manufacturer of Tordon 101 (Dow AgroSciences, owned by the Dow Chemical Company) has claimed Tordon 101 has no effects on animals and insects, in spite of evidence of strong carcinogenic activity of the active ingredient Picloram in studies on rats.

The risk of Parkinson's disease has been shown to increase with occupational exposure to herbicides and pesticides. The herbicide paraquat is suspected to be one such factor.

All commercially sold, organic and nonorganic herbicides must be extensively tested prior to approval for sale and labeling by the Environmental Protection Agency. However, because of the large number of herbicides in use, concern regarding health effects is significant. In addition to health effects caused by herbicides themselves, commercial herbicide mixtures often contain other chemicals, including inactive ingredients, which have negative impacts on human health.

Ecological Effects

Commercial herbicide use generally has negative impacts on bird populations, although the impacts are highly variable and often require field studies to predict accurately. Laboratory studies have at times overestimated negative impacts on birds due to toxicity, predicting serious problems that were not observed in the field. Most observed effects are due not to toxicity, but to habitat changes and the decreases in abundance of species on which birds rely for food or shelter. Herbicide use in silviculture, used to favor certain types of growth following clearcutting, can cause significant drops in bird populations. Even when herbicides which have low toxicity to birds are used, they decrease the abundance of many types of vegetation on which the birds rely. Herbicide use in agriculture in Britain has been linked to a decline in seed-eating bird species which rely on the weeds killed by the herbicides. Heavy use of herbicides in neotropical agricultural areas has been one of many factors implicated in limiting the usefulness of such agricultural land for wintering migratory birds.

Frog populations may be affected negatively by the use of herbicides as well. While some studies have shown that atrazine may be a teratogen, causing demasculinization in male frogs, the U.S. Environmental Protection Agency (EPA) and its independent Scientific Advisory Panel (SAP) examined all available studies on this topic and concluded that "atrazine does not adversely affect amphibian gonadal development based on a review of laboratory and field studies."

Scientific Uncertainty of Full Extent of Herbicide Effects

The health and environmental effects of many herbicides is unknown, and even the scientific community often disagrees on the risk. For example, a 1995 panel of 13 scientists reviewing studies on the carcinogenicity of 2,4-D had divided opinions on the likelihood 2,4-D causes cancer in humans. As of 1992, studies on phenoxy herbicides were too few to accurately assess the risk of many types of cancer from these herbicides, even though evidence was stronger that exposure to these herbicides is associated with increased risk of soft tissue sarcoma and non-Hodgkin lymphoma. Furthermore, there is some suggestion that herbicides can play a role in sex reversal of certain organisms that experience temperature-dependent sex determination, which could theoretically alter sex ratios.

Resistance

Weed resistance to herbicides has become a major concern in crop production worldwide. Resistance to herbicides is often attributed to lack of rotational programmes of herbicides and to continuous applications of herbicides with the same sites of action. Thus, a true understanding of the sites of action of herbicides is essential for strategic planning of herbicide-based weed control.

Plants have developed resistance to atrazine and to ALS-inhibitors, and more recently, to glyphosate herbicides. Marestail is one weed that has developed glyphosate resistance. Glyphosate-resistant weeds are present in the vast majority of soybean, cotton and corn farms in some U.S. states. Weeds that can resist multiple other herbicides are spreading. Few new herbicides are near commercialization, and none with a molecular mode of action for which there is no resistance. Because most herbicides could not kill all weeds, farmers rotated crops and herbicides to stop resistant weeds. During its initial years, glyphosate was not subject to resistance and allowed farmers to reduce the use of rotation.

A family of weeds that includes waterhemp (Amaranthus rudis) is the largest concern. A 2008-9 survey of 144 populations of waterhemp in 41 Missouri counties revealed glyphosate resistance in 69%. Weeds from some 500 sites throughout Iowa in 2011 and 2012 revealed glyphosate resistance in approximately 64% of waterhemp samples. The use of other killers to target "residual" weeds has become common, and may be sufficient to have stopped the spread of resistance From 2005 through 2010 researchers discovered 13 different weed species that had developed resistance to glyphosate. But since then only two more have been discovered. Weeds resistant to multiple herbicides with completely different biological action modes are on the rise. In Missouri, 43% of samples were resistant to two different herbicides; 6% resisted three; and 0.5% resisted four. In Iowa 89% of waterhemp samples resist two or more herbicides, 25% resist three, and 10% resist five.

For southern cotton, herbicide costs has climbed from between $50 and $75 per hectare a few years ago to about $370 per hectare in 2013. Resistance is contributing to a massive shift away from growing cotton; over the past few years, the area planted with cotton has declined by 70% in Arkansas and by 60% in Tennessee. For soybeans in Illinois, costs have risen from about $25 to $160 per hectare.

Dow, Bayer CropScience, Syngenta and Monsanto are all developing seed varieties resistant to herbicides other than glyphosate, which will make it easier for farmers to use alternative weed killers. Even though weeds have already evolved some resistance to those herbicides, Powles says the new seed-and-herbicide combos should work well if used with proper rotation.

Biochemistry of Resistance

Resistance to herbicides can be based on one of the following biochemical mechanisms:

- Target-site resistance: This is due to a reduced (or even lost) ability of the herbicide to bind to its target protein. The effect usually relates to an enzyme with a crucial function in a metabolic pathway, or to a component of an electron-transport system. Target-site resistance may also be caused by an overexpression of the target enzyme (via gene amplification or changes in a gene promoter).

- Non-target-site resistance: This is caused by mechanisms that reduce the amount of herbicidal active compound reaching the target site. One important mechanism is an enhanced metabolic detoxification of the herbicide in the weed, which leads to insufficient amounts of the active substance reaching the target site. A reduced uptake and translocation, or sequestration of the herbicide, may also result in an insufficient herbicide transport to the target site.

- Cross-resistance: In this case, a single resistance mechanism causes resistance to several herbicides. The term target-site cross-resistance is used when the herbicides bind to the same target site, whereas non-target-site cross-resistance is due to a single non-target-site mechanism (e.g., enhanced metabolic detoxification) that entails resistance across herbicides with different sites of action.

- Multiple resistance: In this situation, two or more resistance mechanisms are present within individual plants, or within a plant population.

Resistance Management

Worldwide experience has been that farmers tend to do little to prevent herbicide resistance developing, and only take action when it is a problem on their own farm or neighbor's. Careful observation is important so that any reduction in herbicide efficacy can be detected. This may indicate evolving resistance. It is vital that resistance is detected at an early stage as if it becomes an acute, whole-farm problem, options are more limited and greater expense is almost inevitable. Table 1 lists factors which enable the risk of resistance to be assessed. An essential pre-requisite for confirmation of resistance is a good diagnostic test. Ideally this should be rapid, accurate, cheap and accessible. Many diagnostic tests have been developed, including glasshouse pot assays, petri dish assays and chlorophyll fluorescence. A key component of such tests is that the response of the suspect population to a herbicide can be compared with that of known susceptible and resistant standards under controlled conditions. Most cases of herbicide resistance are a consequence of the repeated use of herbicides, often in association with crop monoculture and reduced cultivation practices. It is necessary, therefore, to modify these practices in order to prevent or delay the onset of resistance or to control existing resistant populations. A key objective should be the reduction in selection pressure. An integrated weed management (IWM) approach is required, in which as many tactics as possible are used to combat weeds. In this way, less reliance is placed on herbicides and so selection pressure should be reduced.

Optimising herbicide input to the economic threshold level should avoid the unnecessary use of herbicides and reduce selection pressure. Herbicides should be used to their greatest potential by ensuring that the timing, dose, application method, soil and climatic conditions are optimal for good activity. In the UK, partially resistant grass weeds such as *Alopecurus myosuroides* (blackgrass) and *Avena* spp. (wild oat) can often be controlled adequately when herbicides are applied at the 2-3 leaf stage, whereas later applications at the 2-3 tiller stage can fail badly. Patch spraying, or applying herbicide to only the badly infested areas of fields, is another means of reducing total herbicide use.

Table 1. Agronomic factors influencing the risk of herbicide resistance development

Factor	Low risk	High risk
Cropping system	Good rotation	Crop monoculture
Cultivation system	Annual ploughing	Continuous minimum tillage
Weed control	Cultural only	Herbicide only
Herbicide use	Many modes of action	Single modes of action
Control in previous years	Excellent	Poor
Weed infestation	Low	High
Resistance in vicinity	Unknown	Common

Approaches to Treating Resistant Weeds

Alternative Herbicides

When resistance is first suspected or confirmed, the efficacy of alternatives is likely to be the first consideration. The use of alternative herbicides which remain effective on resistant populations can be a successful strategy, at least in the short term. The effectiveness of alternative herbicides will be highly dependent on the extent of cross-resistance. If there is resistance to a single group of herbicides, then the use of herbicides from other groups may provide a simple and effective solution, at least in the short term. For example, many triazine-resistant weeds have been readily controlled by the use of alternative herbicides such as dicamba or glyphosate. If resistance extends to more than one herbicide group, then choices are more limited. It should not be assumed that resistance will automatically extend to all herbicides with the same mode of action, although it is wise to assume this until proved otherwise. In many weeds the degree of cross-resistance between the five groups of ALS inhibitors varies considerably. Much will depend on the resistance mechanisms present, and it should not be assumed that these will necessarily be the same in different populations of the same species. These differences are due, at least in part, to the existence of different mutations conferring target site resistance. Consequently, selection for different mutations may result in different patterns of cross-resistance. Enhanced metabolism can affect even closely related herbicides to differing degrees. For example, populations of *Alopecurus myosuroides* (blackgrass) with an enhanced metabolism mechanism show resistance to pendimethalin but not to trifluralin, despite both being dinitroanilines. This is due to differences in the vulnerability of these two herbicides to oxidative metabolism. Consequently, care is needed when trying to predict the efficacy of alternative herbicides.

Mixtures and Sequences

The use of two or more herbicides which have differing modes of action can reduce the selection for resistant genotypes. Ideally, each component in a mixture should:

- Be active at different target sites

- Have a high level of efficacy

- Be detoxified by different biochemical pathways

- Have similar persistence in the soil (if it is a residual herbicide)

- Exert negative cross-resistance

- Synergise the activity of the other component

No mixture is likely to have all these attributes, but the first two listed are the most important. There is a risk that mixtures will select for resistance to both components in the longer term. One practical advantage of sequences of two herbicides compared with mixtures is that a better appraisal of the efficacy of each herbicide component is possible, provided that sufficient time elapses between each application. A disadvantage with sequences is that two separate applications have to be made and it is possible that the later application will be less effective on weeds surviving the first application. If these are resistant, then the second herbicide in the sequence may increase selection for resistant individuals by killing the susceptible plants which were damaged but not killed by the first application, but allowing the larger, less affected, resistant plants to survive. This has been cited as one reason why ALS-resistant *Stellaria media* has evolved in Scotland recently (2000), despite the regular use of a sequence incorporating mecoprop, a herbicide with a different mode of action.

Herbicide Rotations

Rotation of herbicides from different chemical groups in successive years should reduce selection for resistance. This is a key element in most resistance prevention programmes. The value of this approach depends on the extent of cross-resistance, and whether multiple resistance occurs owing to the presence of several different resistance mechanisms. A practical problem can be the lack of awareness by farmers of the different groups of herbicides that exist. In Australia a scheme has been introduced in which identifying letters are included on the product label as a means of enabling farmers to distinguish products with different modes of action.

Farming Practices and Resistance: a Case Study

Herbicide resistance became a critical problem in Australian agriculture, after many Australian sheep farmers began to exclusively grow wheat in their pastures in the 1970s. Introduced varieties of ryegrass, while good for grazing sheep, compete intensely with wheat. Ryegrasses produce so many seeds that, if left unchecked, they can completely choke a field. Herbicides provided excellent control, while reducing soil disrupting because of less need to plough. Within little more than a decade, ryegrass and other weeds began to develop resistance. In response Australian farmers changed methods. By 1983, patches of ryegrass had become immune to Hoegrass, a family of herbicides that inhibit an enzyme called acetyl coenzyme A carboxylase.

Ryegrass populations were large, and had substantial genetic diversity, because farmers had planted many varieties. Ryegrass is cross-pollinated by wind, so genes shuffle frequently. To control its distribution farmers sprayed inexpensive Hoegrass, creating selection pressure. In addition, farmers sometimes diluted the herbicide in order to save money, which allowed some plants to survive application. When resistance appeared farmers turned to a group of herbicides that block acetolactate synthase. Once again, ryegrass in Australia evolved a kind of "cross-resistance" that allowed it to rapidly break down a variety of herbicides. Four classes of herbicides become ineffective within a few years. In 2013 only two herbicide classes, called Photosystem II and long-chain fatty acid inhibitors, were effective against ryegrass.

List of Common Herbicides

Synthetic Herbicides

- 2,4-D is a broadleaf herbicide in the phenoxy group used in turf and no-till field crop production. Now, it is mainly used in a blend with other herbicides to allow lower rates of herbicides to be used; it is the most widely used herbicide in the world, and third most commonly used in the United States. It is an example of synthetic auxin (plant hormone).

- Aminopyralid is a broadleaf herbicide in the pyridine group, used to control weeds on grassland, such as docks, thistles and nettles. It is notorious for its ability to persist in compost.

- Atrazine, a triazine herbicide, is used in corn and sorghum for control of broadleaf weeds and grasses. Still used because of its low cost and because it works well on a broad spectrum of weeds common in the US corn belt, atrazine is commonly used with other herbicides to reduce the overall rate of atrazine and to lower the potential for groundwater contamination; it is a photosystem II inhibitor.

- Clopyralid is a broadleaf herbicide in the pyridine group, used mainly in turf, rangeland, and for control of noxious thistles. Notorious for its ability to persist in compost, it is another example of synthetic auxin.

- Dicamba, a postemergent broadleaf herbicide with some soil activity, is used on turf and field corn. It is another example of a synthetic auxin.

- Glufosinate ammonium, a broad-spectrum contact herbicide, is used to control weeds after the crop emerges or for total vegetation control on land not used for cultivation.

- Fluazifop (Fuselade Forte), a post emergence, foliar absorbed, translocated grass-selective herbicide with little residual action. It is used on a very wide range of broad leaved crops for control of annual and perennial grasses.

- Fluroxypyr, a systemic, selective herbicide, is used for the control of broad-leaved weeds in small grain cereals, maize, pastures, rangeland and turf. It is a synthetic auxin. In cereal growing, fluroxypyr's key importance is control of cleavers, *Galium aparine*. Other key broadleaf weeds are also controlled.

- Glyphosate, a systemic nonselective herbicide, is used in no-till burndown and for weed control in crops genetically modified to resist its effects. It is an example of an EPSPs inhibitor.

- Imazapyr a nonselective herbicide, is used for the control of a broad range of weeds, including terrestrial annual and perennial grasses and broadleaf herbs, woody species, and riparian and emergent aquatic species.

- Imazapic, a selective herbicide for both the pre- and postemergent control of some annual and perennial grasses and some broadleaf weeds, kills plants by inhibiting the production of branched chain amino acids (valine, leucine, and isoleucine), which are necessary for protein synthesis and cell growth.

- Imazamox, an imidazolinone manufactured by BASF for postemergence application that is an acetolactate synthase (ALS) inhibitor. Sold under trade names Raptor, Beyond, and Clearcast.

- Linuron is a nonselective herbicide used in the control of grasses and broadleaf weeds. It works by inhibiting photosynthesis.

- MCPA (2-methyl-4-chlorophenoxyacetic acid) is a phenoxy herbicide selective for broadleaf plants and widely used in cereals and pasture.

- Metolachlor is a pre-emergent herbicide widely used for control of annual grasses in corn and sorghum; it has displaced some of the atrazine in these uses.

- Paraquat is a nonselective contact herbicide used for no-till burndown and in aerial destruction of marijuana and coca plantings. It is more acutely toxic to people than any other herbicide in widespread commercial use.

- Pendimethalin, a pre-emergent herbicide, is widely used to control annual grasses and some broad-leaf weeds in a wide range of crops, including corn, soybeans, wheat, cotton, many tree and vine crops, and many turfgrass species.

- Picloram, a pyridine herbicide, mainly is used to control unwanted trees in pastures and edges of fields. It is another synthetic auxin.

- Sodium chlorate (disused/banned in some countries), a nonselective herbicide, is considered phytotoxic to all green plant parts. It can also kill through root absorption.

- Triclopyr, a systemic, foliar herbicide in the pyridine group, is used to control broadleaf weeds while leaving grasses and conifers unaffected.

- Several sulfonylureas, including Flazasulfuron and Metsulfuron-methyl, which act as ALS inhibitors and in some cases are taken up from the soil via the roots.

Organic herbicides

Recently, the term "organic" has come to imply products used in organic farming. Under this definition, an organic herbicide is one that can be used in a farming enterprise that has been classified as organic. Commercially sold organic herbicides are expensive and may not be affordable for commercial farming. Depending on the application, they may be less effective than synthetic herbicides and are generally used along with cultural and mechanical weed control practices.

Homemade organic herbicides include:

- Corn gluten meal (CGM) is a natural pre-emergence weed control used in turfgrass, which reduces germination of many broadleaf and grass weeds.

- Vinegar is effective for 5–20% solutions of acetic acid, with higher concentrations most effective, but it mainly destroys surface growth, so respraying to treat regrowth is needed. Resistant plants generally succumb when weakened by respraying.

- Steam has been applied commercially, but is now considered uneconomical and inadequate. It controls surface growth but not underground growth and so respraying to treat regrowth of perennials is needed.

- Flame is considered more effective than steam, but suffers from the same difficulties.

- D-limonene (citrus oil) is a natural degreasing agent that strips the waxy skin or cuticle from weeds, causing dehydration and ultimately death.

- Saltwater or salt applied in appropriate strengths to the rootzone will kill most plants.

- Monocerin produced by certain fungi will kill certain weeds such as Johnson grass.

Of Historical Interest and Other

- 2,4,5-Trichlorophenoxyacetic acid (2,4,5-T) was a widely used broadleaf herbicide until being phased out starting in the late 1970s. While 2,4,5-T itself is of only moderate toxicity, the manufacturing process for 2,4,5-T contaminates this chemical with trace amounts of 2,3,7,8-tetrachlorodibenzo-p-dioxin (TCDD). TCDD is extremely toxic to humans. With proper temperature control during production of 2,4,5-T, TCDD levels can be held to about .005 ppm. Before the TCDD risk was well understood, early production facilities lacked proper temperature controls. Individual batches tested later were found to have as much as 60 ppm of TCDD. 2,4,5-T was withdrawn from use in the USA in 1983, at a time of heightened public sensitivity about chemical hazards in the environment. Public concern about dioxins was high, and production and use of other (non-herbicide) chemicals potentially containing TCDD contamination was also withdrawn. These included pentachlorophenol (a wood preservative) and PCBs (mainly used as stabilizing agents in transformer oil). Some feel that the 2,4,5-T withdrawal was not based on sound science. 2,4,5-T has since largely been replaced by dicamba and triclopyr.

- Agent Orange was a herbicide blend used by the British military during the Malayan Emergency and the U.S. military during the Vietnam War between January 1965 and April 1970 as a defoliant. It was a 50/50 mixture of the n-butyl esters of 2,4,5-T and 2,4-D. Because of TCDD contamination in the 2,4,5-T component, it has been blamed for serious illnesses in many people who were exposed to it. However, research on populations exposed to its dioxin contaminant have been inconsistent and inconclusive.

- Diesel, and other heavy oil derivatives, are known to be informally used at times, but are usually banned for this purpose.

Rodenticide

Rodenticides, colloquially rat poison, are typically non-specific pest control chemicals made and sold for the purpose of killing rodents.

Some rodenticides are lethal after one exposure while others require more than one. Rodents are disinclined to gorge on an unknown food (perhaps reflecting an adaptation to their inability to vomit), preferring to sample, wait and observe whether it makes them or other rats sick. This phenomenon of bait shyness or poison shyness is the rationale for poisons that kill only after multiple doses.

Besides being directly toxic to the mammals that ingest them, including dogs, cats, and humans, many rodenticides present a secondary poisoning risk to animals that hunt or scavenge the dead corpses of rats.

Chemical Preparations

Anticoagulants

Anticoagulants are defined as chronic (death occurs one to two weeks after ingestion of the lethal dose, rarely sooner), single-dose (second generation) or multiple-dose (first generation) rodenticides, acting by effective blocking of the vitamin K cycle, resulting in inability to produce essential blood-clotting factors — mainly coagulation factors II (prothrombin) and VII (proconvertin).

In addition to this specific metabolic disruption, massive toxic doses of 4-hydroxycoumarin, 4-thiochromenone and indandione anticoagulants cause damage to tiny blood vessels (capillaries), increasing their permeability, causing diffuse internal bleeding. These effects are gradual, developing over several days. In the final phase of the intoxication, the exhausted rodent collapses due to hemorrhagic shock or severe anemia and dies calmly. The question of whether the use of these rodenticides can be considered humane has been raised.

The main benefit of anticoagulants over other poisons is that the time taken for the poison to induce death means that the rats do not associate the damage with their feeding habits.

- First generation rodenticidal anticoagulants generally have shorter elimination half-lives, require higher concentrations (usually between 0.005% and 0.1%) and consecutive intake over days in order to accumulate the lethal dose, and are less toxic than second generation agents.

- Second generation agents are far more toxic than first generation. They are generally applied in lower concentrations in baits — usually on the order of 0.001% to 0.005% — are lethal after a single ingestion of bait and are also effective against strains of rodents that became resistant to first generation anticoagulants; thus, the second generation anticoagulants are sometimes referred to as "superwarfarins".

Class	Examples
Coumarins/4-hydroxycoumarins	• First generation: warfarin, coumatetralyl • Second generation: difenacoum, brodifacoum, flocoumafen and bromadiolone.

1,3-indandiones	diphacinone, chlorophacinone, pindone These are harder to group by generation. According to some sources, the indandiones are considered second generation. However, according to the U.S. Environmental Protection Agency, examples of first generation agents include chlorophacinone and diphacinone.
4-thiochromenones	Difethialone is considered a second generation anticoagulant rodenticide .
Indirect	Sometimes, anticoagulant rodenticides are potentiated by an antibiotic or bacteriostatic agent, most commonly sulfaquinoxaline. The aim of this association is that the antibiotic suppresses intestinal symbiotic microflora, which are a source of vitamin K. Diminished production of vitamin K by the intestinal microflora contributes to the action of anticoagulants. Added vitamin D also has a synergistic effect with anticoagulants.

Vitamin K_1 has been suggested, and successfully used, as antidote for pets or humans accidentally or intentionally exposed to anticoagulant poisons. Some of these poisons act by inhibiting liver functions and in advanced stages of poisoning, several blood-clotting factors are absent, and the volume of circulating blood is diminished, so that a blood transfusion (optionally with the clotting factors present) can save a person who has been poisoned, an advantage over some older poisons.

Metal Phosphides

Rat poison vendor's stall at a market in Linxia City, China

Metal phosphides have been used as a means of killing rodents and are considered single-dose fast acting rodenticides (death occurs commonly within 1–3 days after single bait ingestion). A bait consisting of food and a phosphide (usually zinc phosphide) is left where the rodents can eat it. The acid in the digestive system of the rodent reacts with the phosphide to generate the toxic phosphine gas. This method of vermin control has possible use in places where rodents are resistant to some of the anticoagulants, particularly for control of house and field mice; zinc phosphide baits are also cheaper than most second-generation anticoagulants, so that sometimes, in the case of large infestation by rodents, their population is initially reduced by copious amounts of zinc phosphide bait applied, and the rest of population that survived the initial fast-acting poison is then eradicated by prolonged feeding on anticoagulant bait. Inversely, the individual rodents, that survived anticoagulant bait poisoning (rest population) can be eradicated by pre-baiting them with nontoxic bait for a week or two (this is important to overcome bait shyness, and to get rodents used to feeding in specific areas by specific food, especially in eradicating rats) and subsequently applying poisoned bait of the same sort as used for pre-baiting until all consumption of the bait ceases (usually within 2–4 days). These methods of alternating rodenticides with different modes of action gives actual or almost 100% eradications of the rodent population in the area, if the acceptance/palatability of baits are good (i.e., rodents feed on it readily).

Zinc phosphide is typically added to rodent baits in a concentration of 0.75% to 2.0%. The baits have strong, pungent garlic-like odor due to the phosphine liberated by hydrolysis. The odor attracts (or, at least, does not repel) rodents, but has an repulsive effect on other mammals. Birds, notably wild turkeys, are not sensitive to the smell, and will feed on the bait, and thus become collateral damage.

The tablets or pellets (usually aluminium, calcium or magnesium phosphide for fumigation/gassing) may also contain other chemicals which evolve ammonia, which helps to reduce the potential for spontaneous combustion or explosion of the phosphine gas.

Metal phosphides do not accumulate in the tissues of poisoned animals, so the risk of secondary poisoning is low.

Before the advent of anticoagulants, phosphides were the favored kind of rat poison. During World War II, they came into use in United States because of shortage of strychnine due to the Japanese occupation of the territories where the strychnine tree is grown. Phosphides are rather fast-acting rat poisons, resulting in the rats dying usually in open areas, instead of in the affected buildings.

Phosphides used as rodenticides include:

- aluminium phosphide (fumigant only)

- calcium phosphide (fumigant only)

- magnesium phosphide (fumigant only)

- zinc phosphide (bait only)

Hypercalcemia

Calciferols (vitamins D), cholecalciferol (vitamin D_3) and ergocalciferol (vitamin D_2) are used as rodenticides. They are toxic to rodents for the same reason they are important to humans: they affect calcium and phosphate homeostasis in the body. Vitamins

D are essential in minute quantities (few IUs per kilogram body weight daily, only a fraction of a milligram), and like most fat soluble vitamins, they are toxic in larger doses, causing hypervitaminosis. If the poisoning is severe enough (that is, if the dose of the toxin is high enough), it leads to death. In rodents that consume the rodenticidal bait, it causes hypercalcemia, raising the calcium level, mainly by increasing calcium absorption from food, mobilising bone-matrix-fixed calcium into ionised form (mainly monohydrogencarbonate calcium cation, partially bound to plasma proteins, $[CaH-CO_3]^+$), which circulates dissolved in the blood plasma. After ingestion of a lethal dose, the free calcium levels are raised sufficiently that blood vessels, kidneys, the stomach wall and lungs are mineralised/calcificated (formation of calcificates, crystals of calcium salts/complexes in the tissues, damaging them), leading further to heart problems (myocardial tissue is sensitive to variations of free calcium levels, affecting both myocardial contractibility and excitation propagation between atrias and ventriculas), bleeding (due to capillary damage) and possibly kidney failure. It is considered to be single-dose, cumulative (depending on concentration used; the common 0.075% bait concentration is lethal to most rodents after a single intake of larger portions of the bait) or sub-chronic (death occurring usually within days to one week after ingestion of the bait). Applied concentrations are 0.075% cholecalciferol and 0.1% ergocalciferol when used alone, wihich can kill a rodent or a rat.

There is an important feature of calciferols toxicology, that they are synergistic with anticoagulant toxicants, that means, that mixtures of anticoagulants and calciferols in same bait are more toxic than a sum of toxicities of the anticoagulant and the calciferol in the bait, so that a massive hypercalcemic effect can be achieved by a substantially lower calciferol content in the bait, and vice versa, a more pronounced anticoagulant/hemorrhagic effects are observed if the calciferol is present. This synergism is mostly used in calciferol low concentration baits, because effective concentrations of calciferols are more expensive than effective concentrations of most anticoagulants.

The first application of a calciferol in rodenticidal bait was in the Sorex product Sorexa D (with a different formula than today's Sorexa D), back in the early 1970s, which contained 0.025% warfarin and 0.1% ergocalciferol. Today, Sorexa CD contains a 0.0025% difenacoum and 0.075% cholecalciferol combination. Numerous other brand products containing either 0.075-0.1% calciferols (e.g. Quintox) alone or alongside an anticoagulant are marketed.

The Merck Veterinary Manual states the following:

Although this rodenticide [cholecalciferol] was introduced with claims that it was less toxic to nontarget species than to rodents, clinical experience has shown that rodenticides containing cholecalciferol are a significant health threat to dogs and cats. Cholecalciferol produces hypercalcemia, which results in systemic calcification of soft tissue, leading to renal failure, cardiac abnormalities, hypertension, CNS depression and GI upset. Signs generally develop within 18-36 hours of ingestion and can include depression, anorexia, polyuria and polydipsia. As serum calcium concentrations increase, clinical signs become more severe. ... GI smooth muscle excitability decreases and is manifest by anorexia, vomiting and constipation. ... Loss of renal concentrating ability is a direct result of hypercalcemia. As hypercalcemia persists, mineralization of the kidneys results in progressive renal insufficiency."

Additional anticoagulant renders the bait more toxic to pets as well as human. Upon single ingestion, solely calciferol-based baits are considered generally safer to birds than second generation anticoagulants or acute toxicants. A specific antidote for calciferol intoxication is calcitonin, a hormone that lowers the blood levels of calcium. The therapy with commercially available calcitonin preparations is, however, expensive.

Other

Civilian Public Service worker distributes rat poison for typhus control in Gulfport, Mississippi, ca. 1945.

Other chemical poisons include:

- ANTU (α-naphthylthiourea; specific against Brown rat, *Rattus norvegicus*)
- Arsenic trioxide
- Barium carbonate
- Chloralose (a narcotic prodrug)
- Crimidine (inhibits metabolism of vitamin B_6)
- 1,3-Difluoro-2-propanol ("Gliftor")
- Endrin (organochlorine insecticide, used in the past for extermination of voles in fields)
- Fluoroacetamide ("1081")
- Phosacetim (a delayed-action organophosphate)
- White phosphorus

- Pyrinuron (an urea derivative)

- Scilliroside

- Sodium fluoroacetate ("1080")

- Strychnine (A naturally occurring convulsant and stimulant)

- Tetramethylenedisulfotetramine ("tetramine")

- Thallium sulfate

- Nitrophenols like bromethalin and 2,4-dinitrophenol (cause high fever and brain swelling, no known antidote)

- Zyklon B/Uragan D2 (hydrogen cyanide gas absorbed in an inert carrier)

Combinations

In some countries, fixed three-component rodenticides, i.e., anticoagulant + antibiotic + vitamin D, are used. Associations of a second-generation anticoagulant with an antibiotic and/or vitamin D are considered to be effective even against most resistant strains of rodents, though some second generation anticoagulants (namely brodifacoum and difethialone), in bait concentrations of 0.0025% to 0.005% are so toxic that resistance is unknown, and even rodents resistant to other rodenticides are reliably exterminated by application of these most toxic anticoagulants.

Alternatives

More environmentally-safe preparations, such as powdered corn cob, have been developed and were approved in the EU and patented in the US in 2013. These preparations rely on dehydration to cause death.

Non-Target Issues

Secondary Poisoning and Risks to Wildlife

One of the potential problems when using rodenticides is that dead or weakened rodents may be eaten by other wildlife, either predators or scavengers. Members of the public deploying rodenticides may not be aware of this or may not follow the product's instructions closely enough.

The faster a rodenticide acts, the more critical this problem may be. For the fast-acting rodenticide bromethalin, for example, there is no diagnostic test or antidote.

This has led environmental researchers to conclude that low strength, long duration rodenticides (generally first generation anticoagulants) are the best balance between maximum effect and minimum risk.

Proposed US Legislation Change

In 2008, after assessing human health and ecological effects, as well as benefits, the US Environmental Protection Agency (EPA) announced measures to reduce risks associated with ten rodenticides. New restrictions by sale and distribution restrictions, minimum package size requirements, use site restriction, and tamper resistant products would have taken effect in 2011. The regulations were delayed pending a legal challenge by manufacturer Reckitt-Benkiser.

Disinfectant

Disinfectants are antimicrobial agents that are applied to non-living objects to destroy microorganisms that are living on the objects. Disinfection does not necessarily kill all microorganisms, especially resistant bacterial spores; it is less effective than sterilization, which is an extreme physical and/or chemical process that kills all types of life. Disinfectants are different from other antimicrobial agents such as antibiotics, which destroy microorganisms within the body, and antiseptics, which destroy microorganisms on living tissue. Disinfectants are also different from biocides — the latter are intended to destroy all forms of life, not just microorganisms. Disinfectants work by destroying the cell wall of microbes or interfering with the metabolism.

Sanitizers are substances that simultaneously clean and disinfect. Disinfectants are frequently used in hospitals, dental surgeries, kitchens, and bathrooms to kill infectious organisms.

Bacterial endospores are most resistant to disinfectants, but some viruses and bacteria also possess some tolerance.

In wastewater treatment, a disinfection step with chlorine, ultra-violet (UV) radiation or ozonation can be included as tertiary treatment to remove pathogens from wastewater, for example if it is to be reused to irrigate golf courses. An alternative term used in the sanitation sector for disinfection of waste streams, sewage sludge or fecal sludge is sanitisation or sanitization.

Properties

A perfect disinfectant would also offer complete and full microbiological sterilisation, without harming humans and useful form of life, be inexpensive, and noncorrosive. However, most disinfectants are also, by nature, potentially harmful (even toxic) to humans or animals. Most modern household disinfectants contain Bitrex, an exceptionally bitter substance added to discourage ingestion, as a safety measure. Those that are used indoors should never be mixed with other cleaning products as chemical reactions can occur. The choice of disinfectant to be used depends on the particular situation. Some disinfectants have a wide spectrum (kill many different types of microorganisms), while others kill a smaller range of disease-causing organisms but are preferred for other properties (they may be non-corrosive, non-toxic, or inexpensive). There are arguments for creating or maintaining conditions that are not conducive to bacterial survival and multiplication, rather than attempting to kill them with chemicals. Bacteria can increase in number very quickly, which enables them to evolve rapidly. Should some bacteria survive a chemical attack, they give rise to new generations composed completely of bacteria that have resistance to the particular chemical used. Under a sustained chemical attack, the surviving bacteria in successive generations are increasingly resistant to the chemical used, and ultimately the chemical is rendered ineffective. For this reason, some question the wisdom of impregnating cloths, cutting boards and worktops in the home with bactericidal chemicals.

Types

Air Disinfectants

Air disinfectants are typically chemical substances capable of disinfecting microorganisms suspended in the air. Disinfectants are generally assumed to be limited to use on surfaces, but that is not the case. In 1928, a study found that airborne microorganisms could be killed using mists of dilute bleach. An air disinfectant must be dispersed either as an aerosol or vapour at a sufficient concentration in the air to cause the number of viable infectious microorganisms to be significantly reduced.

In the 1940s and early 1950s, further studies showed inactivation of diverse bacteria, influenza virus, and *Penicillium chrysogenum* (previously *P. notatum*) mold fungus using various glycols, principally propylene glycol and triethylene glycol. In principle, these chemical substances are ideal air disinfectants because they have both high lethality to microorganisms and low mammalian toxicity.

Although glycols are effective air disinfectants in controlled laboratory environments, it is more difficult to use them effectively in real-world environments because the disinfection of air is sensitive to continuous action. Continuous action in real-world environments with outside air exchanges at door, HVAC, and window interfaces, and in the presence of materials that adsorb and remove glycols from the air, poses engineering challenges that are not critical for surface disinfection. The engineering challenge associated with creating a sufficient concentration of the glycol vapours in the air have not to date been sufficiently addressed.

Alcohols

Alcohol and alcohol plus Quaternary ammonium cation based compounds comprise a class of proven surface sanitizers and disinfectants approved by the EPA and the Centers for Disease Control for use as a hospital grade disinfectant. Alcohols are most effective when combined with distilled water to facilitate diffusion through the cell membrane; 100% alcohol typically denatures only external membrane proteins. A mixture of 70% ethanol or isopropanol diluted in water is effective against a wide spectrum of bacteria, though higher concentrations are often needed to disinfect wet surfaces. Additionally, high-concentration mixtures (such as 80% ethanol + 5% isopropanol) are required to effectively inactivate lipid-enveloped viruses (such as HIV, hepatitis B, and hepatitis C). The efficacy of alcohol is enhanced when in solution with the wetting agent dodecanoic acid (coconut soap). The synergistic effect of 29.4% ethanol with dodecanoic acid is effective against a broad spectrum of bacteria, fungi, and viruses. Further testing is being performed against Clostridium difficile (C.Diff) spores with higher concentrations of ethanol and dodecanoic acid, which proved effective with a contact time of ten minutes.

Aldehydes

Aldehydes, such as formaldehyde and glutaraldehyde, have a wide microbiocidal activity and are sporicidal and fungicidal. They are partly inactivated by organic matter and have slight residual activity.

Some bacteria have developed resistance to glutaraldehyde, and it has been found that glutaraldehyde can cause asthma and other health hazards, hence ortho-phthalaldehyde is replacing glutaraldehyde.

Oxidizing Agents

Oxidizing agents act by oxidizing the cell membrane of microorganisms, which results in a loss of structure and leads to cell lysis and death. A large number of disinfectants operate in this way. Chlorine and oxygen are strong oxidizers, so their compounds figure heavily here.

- Sodium hypochlorite is very commonly used. Common household bleach is a sodium hypochlorite solution and is used in the home to disinfect drains, toilets, and other surfaces. In more dilute form, it is used in swimming pools, and in still more dilute form, it is used in drinking water. When pools and drinking water are said to be chlorinated, it is actually sodium hypochlorite or a related compound—not pure chlorine—that is being used. Chlorine partly reacts with proteinaceous liquids such as blood to form non-oxidizing N-chloro compounds, and thus higher concentrations must be used if disinfecting surfaces after blood spills. Commercial solutions with higher concentrations contain substantial amounts of sodium hydroxide for stabilization of the concentrated hypochlorite, which would otherwise decompose to chlorine, but the solutions are strongly basic as a result.

- Other hypochlorites such as calcium hypochlorite are also used, especially as a swimming pool additive. Hypochlorites yield an aqueous solution of hypochlorous acid that is the true disinfectant. Hypobromite solutions are also sometimes used.

- Electrolyzed water or "Anolyte" is an oxidizing, acidic hypochlorite solution made by electrolysis of sodium chloride into sodium hypochlorite and hypochlorous acid. Anolyte has an oxidation-reduction potential of +600 to +1200 mV and a typical pH range of 3.5––8.5, but the most potent solution is produced at a controlled pH 5.0–6.3 where the predominant oxychlorine species is hypochlorous acid.

- Chloramine is often used in drinking water treatment.

- Chloramine-T is antibacterial even after the chlorine has been spent, since the parent compound is a sulfonamide antibiotic.

- Chlorine dioxide is used as an advanced disinfectant for drinking water to reduce waterborne diseases. In certain parts of the world, it has largely replaced chlorine because it forms fewer byproducts. Sodium chlorite, sodium chlorate, and potassium chlorate are used as precursors for generating chlorine dioxide.

- Hydrogen peroxide is used in hospitals to disinfect surfaces and it is used in solution alone or in combination with other chemicals as a high level disinfectant. Hydrogen peroxide is sometimes mixed with colloidal silver. It is often preferred because it causes far fewer allergic reactions than alternative disinfectants. Also used in the food packaging industry to disinfect foil containers. A 3% solution is also used as an antiseptic.

- Hydrogen peroxide vapor is used as a medical sterilant and as room disinfectant. Hydrogen peroxide has the advantage that it decomposes to form oxygen and water thus leaving no long term residues, but hydrogen peroxide as with most other strong oxidants is hazardous, and solutions are a primary irritant. The vapor is hazardous to the respiratory system and eyes and consequently the OSHA permissible exposure limit is 1 ppm (29 CFR 1910.1000 Table Z-1) calculated as an eight-hour time weighted average and the NIOSH immediately dangerous to life and health limit is 75 ppm. Therefore, engineering controls, personal protective equipment, gas monitoring etc. should be employed where high concentrations of hydrogen peroxide are used in the workplace. Vaporized hydrogen peroxide is one of the chemicals approved for decontamination of anthrax spores from contaminated buildings, such as occurred during the 2001 anthrax attacks in the U.S. It has also been shown to be effective in removing exotic animal viruses, such as avian influenza and Newcastle disease from equipment and surfaces.

- The antimicrobial action of hydrogen peroxide can be enhanced by surfactants and organic acids. The resulting chemistry is known as Accelerated Hydrogen Peroxide. A 2% solution, stabilized for extended use, achieves high-level disinfection in 5 minutes, and is suitable for disinfecting medical equipment made from hard plastic, such as in endoscopes. The evidence available suggests that products based on Accelerated Hydrogen Peroxide, apart from being good germicides, are safer for humans and benign to the environment.

- Iodine is usually dissolved in an organic solvent or as Lugol's iodine solution. It is used in the poultry industry. It is added to the birds' drinking water. In human and veterinary medicine, iodine products are widely used to prepare incision sites prior to surgery. Although it increases both scar tissue formation and healing time, tincture of iodine is used as an antiseptic for skin cuts and scrapes, and remains among the most effective antiseptics known. Also used as an iodophor

- Ozone is a gas used for disinfecting water, laundry, foods, air, and surfaces. It is chemically aggressive and destroys many organic compounds, resulting in rapid decolorization and deodorization in addition to disinfection. Ozone decomposes relatively quickly. However, due to this characteristic of ozone, tap water chlorination cannot be entirely replaced by ozonation, as the ozone would decompose already in the water piping. Instead, it is used to remove the bulk of oxidizable matter from the water, which would produce small amounts of organochlorides if treated with chlorine only. Regardless, ozone has a very wide range of applications from municipal to industrial water treatment due to its powerful reactivity.

- Peracetic acid is a disinfectant produced by reacting hydrogen peroxide with acetic acid. It is broadly effective against microorganisms and is not deactivated by catalase and peroxidase, the enzymes that break down hydrogen peroxide. It also breaks down to food safe and environmentally friendly residues (acetic acid and hydrogen peroxide), and therefore can be used in non-rinse applications. It can be used over a wide temperature range (0-40 °C), wide pH range (3.0-7.5), in clean-in-place (CIP) processes, in hard water conditions, and is not affected by protein residues.

- Performic acid is the simplest and most powerful perorganic acid. Formed from the reaction of hydrogen peroxide and formic acid, it reacts more rapidly and powerfully than peracetic acid before breaking down to water and carbon dioxide.

- Potassium permanganate ($KMnO_4$) is a purplish-black crystalline powder that colours everything it touches, through a strong oxidising action. This includes staining "stainless" steel, which somehow limits its use and makes it necessary to use plastic or glass containers. It is used to disinfect aquariums and is also widely used in community swimming pools to disinfect ones feet before entering the pool. Typically, a large shallow basin of $KMnO_4$/water solution is kept near the pool ladder. Participants are required to step in the basin and then go into the pool. Additionally, it is widely used to disinfect community water ponds and wells in tropical countries, as well as to disinfect the mouth before pulling out teeth. It can be applied to wounds in dilute solution.

- Potassium peroxymonosulfate, the principal ingredient in Virkon, is a wide-spectrum disinfectant used in laboratories. Virkon kills bacteria, viruses, and fungi. It is used as a 1% solution in water, and keeps for one week once it is made up. It is expensive, but very effective, its pink colour fades as it is used up so it is possible to see at a glance if it is still fresh.

Phenolics

Phenolics are active ingredients in some household disinfectants. They are also found in some mouthwashes and in disinfectant soap and handwashes. Phenols are toxic to cats and newborn humans

- Phenol is probably the oldest known disinfectant as it was first used by Lister, when it was called carbolic acid. It is rather corrosive to the skin and sometimes toxic to sensitive people. Impure preparations of phenol were originally made from coal tar, and these contained low concentrations of other aromatic hydrocarbons including benzene, which is an IARC Group 1 carcinogen.

- o-Phenylphenol is often used instead of phenol, since it is somewhat less corrosive.

- Chloroxylenol is the principal ingredient in Dettol, a household disinfectant and antiseptic.

- Hexachlorophene is a phenolic that was once used as a germicidal additive to some household products but was banned due to suspected harmful effects.

- Thymol, derived from the herb thyme, is the active ingredient in some "broad spectrum" disinfectants that bears ecological claims.

- Amylmetacresol is found in Strepsils, a throat disinfectant.

- Although not a phenol, 2,4-dichlorobenzyl alcohol has similar effects as phenols, but it cannot inactivate viruses.

Quaternary Ammonium Compounds

Quaternary ammonium compounds ("quats"), such as benzalkonium chloride, are a large group of related compounds. Some concentrated formulations have been shown to be effective low-level disinfectants. Quaternary Ammonia at or above 200ppm plus Alcohol solutions exhibit efficacy against difficult to kill non-enveloped viruses such as norovirus, rotavirus, or polio virus. Newer synergous, low-alcohol formulations are highly effective broad-spectrum disinfectants with quick contact times (3–5 minutes) against bacteria, enveloped viruses, pathogenic fungi, and mycobacteria. Quats are biocides that also kill algae and are used as an additive in large-scale industrial water systems to minimize undesired biological growth.

Silver

Silver has antimicrobial properties, but compounds suitable for disinfection are usually unstable and have a limited shelf-life. Silver dihydrogen citrate (SDC) is a chelated form of silver that maintains its stability. SDC kills microorganisms by two modes of action: 1) the silver ion deactivates structural and metabolic membrane proteins, leading to microbial death; 2) the microbes view SDC as a food source, allowing the silver ion to enter the microbe. Once inside the organism, the silver ion denatures the DNA, which halts the microbe's ability to replicate, leading to its death. This dual action makes SDC highly and quickly effective against a broad spectrum of microbes. SDC is non-toxic, non-caustic, colorless, odorless, and tasteless, and does not produce toxic fumes. SDC is non-toxic to humans and animals: the United States Environmental Protection Agency classifies it into the lowest toxicity category for disinfectants, category IV.

A meta-analysis of 26 studies by the Cochrane Collaboration found that, most were small and of poor quality, and that there was not enough evidence to support the use of silver-containing dressings or creams, as generally these treatments did not promote wound healing or prevent wound infections. Some evidence suggested that silver sulphadiazine had no effect on infection, and actually slowed healing.

Copper Alloy Surfaces

Copper-alloy surfaces have natural intrinsic properties to destroy a wide range of microorganisms (e.g., *E. coli* O157:H7, methicillin-resistant *Staphylococcus aureus* (MRSA), *Staphylococcus*, *Clostridium difficile*, influenza A virus, adenovirus, and fungi). In addition, extensive tests on E. coli O157:H7, methicillin-resistant *Staphylococcus aureus* (MRSA), *Staphylococcus*, *Enterobacter aerogenes*, and *Pseudomonas aeruginosa* sanctioned by the United States Environmental Protection Agency (EPA) using Good Laboratory Practices found that when cleaned regularly, some 355 different copper alloy surfaces:

- Continuously reduce bacterial contamination, achieving 99.9% reduction within two hours of exposure;

- Kill greater than 99.9% of Gram-negative and Gram-positive bacteria within two hours of exposure;

- Deliver continuous and ongoing antibacterial action, remaining effective in killing greater than 99.9% of bacteria within two hours;

- Kill greater than 99.9% of bacteria within two hours, and continue to kill 99% of bacteria even after repeated contamination;

- Help inhibit the buildup and growth of bacteria within two hours of exposure between routine cleaning and sanitizing steps.

These copper alloys were granted EPA registrations as "antimicrobial materials with public health benefits," which allows manufacturers to legally make claims regarding the positive public health benefits of products made with registered antimicrobial copper alloys. EPA has approved a long list of antimicrobial copper products made from these alloys, such as bedrails, handrails, over-bed tables, sinks, faucets, door knobs, toilet hardware, computer keyboards, health club equipment, shopping cart handles, etc. (for a comprehensive list of products, Antimicrobial copper-alloy touch surfaces#Approved products). Antimicrobial copper alloy products are now being installed in healthcare facilities in the U.K., Ireland, Japan, Korea, France, Denmark, and Brazil and in the subway transit system in Santiago, Chile, where copper-zinc alloy handrails will be installed in some 30 stations between 2011 and 2014.

Thymol-Based Disinfectant

Thymol, a phenolic chemical found in thyme, can be as effective as bleach in terms of disinfecting as both are considered an intermediate level disinfectant. Thyme essential oils have bacteriostatic activity against a variety of microorganisms, including *E. coli* and *S. aureus*.

Other

The biguanide polymer polyaminopropyl biguanide is specifically bactericidal at very low concentrations (10 mg/l). It has a unique method of action: The polymer strands are incorporated into the bacterial cell wall, which disrupts the membrane and reduces its permeability, which has a lethal effect to bacteria. It is also known to bind to bacterial DNA, alter its transcription, and cause lethal DNA damage. It has very low toxicity to higher organisms such as human cells, which have more complex and protective membranes.

Ultraviolet germicidal irradiation is the use of high-intensity shortwave ultraviolet light for disinfecting smooth surfaces such as dental tools, but not porous materials that are opaque to the light such as wood or foam. Ultraviolet light is also used for municipal water treatment. Ultraviolet light fixtures are often present in microbiology labs, and are activated only when there are no occupants in a room (e.g., at night).

Common sodium bicarbonate ($NaHCO_3$) has antifungal properties, and some antiviral and antibacterial properties, though those are too weak to be effective at a home environment.

Lactic acid is a registered disinfectant. Due to its natural and environmental profile, it has gained importance in the market.

Measurements of Effectiveness

One way to compare disinfectants is to compare how well they do against a known disinfectant and rate them accordingly. Phenol is the standard, and the corresponding rating system is called the "Phenol coefficient". The disinfectant to be tested is compared with phenol on a standard microbe (usually *Salmonella typhi* or *Staphylococcus aureus*). Disinfectants that are more effective than phenol have a coefficient > 1. Those that are less effective have a coefficient < 1.

The standard European approach for disinfectant validation consists of a basic suspension test, a quantitative suspension test (with low and high levels of organic material added to act as 'interfering substances') and a two part simulated-use surface test.

A less specific measurement of effectiveness is the United States Environmental Protection Agency (EPA) classification into either *high, intermediate* or *low* levels of disinfection. "High-level disinfection kills all organisms, except high levels of bacterial spores" and is done with a chemical germicide marketed as a sterilant by the U.S. Food and Drug Administration (FDA). "Intermediate-level disinfection kills mycobacteria, most viruses, and bacteria with a chemical germicide registered as a 'tuberculocide' by the Environmental Protection Agency. Low-level disinfection kills some viruses and bacteria with a chemical germicide registered as a hospital disinfectant by the EPA."

An alternative assessment is to measure the Minimum inhibitory concentrations (MICs) of disinfectants against selected (and representative) microbial species, such as through the use of microbroth dilution testing.

Home Disinfectants

By far the most cost-effective home disinfectant is the commonly used chlorine bleach (a 5% solution of sodium hypochlorite), which is effective against most common pathogens, including difficult organisms such as tuberculosis (mycobacterium tuberculosis), hepatitis B and C, fungi, and antibiotic-resistant strains of staphylococcus and enterococcus. It even has some disinfectant action against parasitic organisms.

Positives are that it kills the widest range of pathogens of any inexpensive disinfectant, is extremely powerful against viruses and bacteria at room temperature, is commonly available and inexpensive, and breaks down quickly into harmless components (primarily table salt and oxygen).

Doors at the Hong Kong Museum of History with signage stating that the doors are disinfected frequently.

Negatives are that it is caustic to the skin, lungs, and eyes (especially at higher concentrations); like many common disinfectants, it degrades in the presence of organic substances; it has a strong odor; it is not effective against *Giardia lamblia* and *Cryptosporidium*; and extreme caution must be taken not to combine it with ammonia or any acid (such as vinegar), as this can cause noxious gases to be formed. The best practice is not to add anything to household bleach except water.

To use chlorine bleach effectively, the surface or item to be disinfected must be clean. In the bathroom or when cleaning after pets, special caution must be taken to wipe up urine first, before applying chlorine, to avoid reaction with the ammonia in urine, causing toxic gas by-products. A 1-to-20 solution in water is effective simply by being wiped on and left to dry. The user should wear rubber gloves and, in tight airless spaces, goggles. If parasitic organisms are suspected, it should be applied at 1-to-1 concentration, or even undiluted. Extreme caution must be taken to avoid contact with eyes and mucous membranes. Protective goggles and good ventilation are mandatory when applying concentrated bleach.

Commercial bleach tends to lose strength over time, whenever the container is opened. Old containers of partially used bleach may no longer have the labeled concentration.

Where one does not want to risk the corrosive effects of bleach, alcohol-based disinfectants are reasonably inexpensive and quite safe. The great drawback to them is their rapid evaporation; sometimes effective disinfection can be obtained only by immersing an object in the alcohol.

The use of some antimicrobials such as triclosan, in particular in the uncontrolled home environment, is controversial because it may lead to the germs becoming resistant. Chlorine bleach and alcohol do not cause resistance because they are so completely lethal, in a very direct physical way.

Insect Growth Regulator

An insect growth regulator (IGR) is a substance (chemical) that inhibits the life cycle of an insect. IGRs are typically used as insecticides to control populations of harmful pests, such as cockroaches or fleas.

Advantages

Many IGRs are labeled "reduced risk" by the Environmental Protection Agency, meaning that they target juvenile harmful insect populations while causing less detrimental effects to beneficial insects. Many beekeepers have reported IGR's negatively affecting brood and young bees . Unlike classic insecticides, IGRs do not affect an insect's nervous system and are thus more worker-friendly within closed environments. IGRs are also more compatible with pest management systems that use biological controls. In addition, while insects can become resistant to insecticides, they are less likely to become resistant to IGRs.

How IGRs Work

As an insect grows, it undergoes a process called molting, where it grows a new exoskeleton under its old one and then sheds to allow the new one to swell to a new size and harden. IGRs prevent an insect from reaching maturity by interfering with the molting process. This in turn curbs infestations because immature insects cannot reproduce. Because IGRs work by interfering with an insect's molting process, they take longer to kill than traditional insecticides. Death typically occurs within 3 to 10 days, depending on the product, the insect's life stage when the product is applied and how quickly the insect develops. Some IGRs cause insects to stop feeding long before they die.

Hormonal IGRs

Hormonal IGRs typically work by mimicking or inhibiting the juvenile hormone (JH), one of the two major hormones involved in insect molting. IGRs can also inhibit the other hormone, ecdysone, large peaks of which trigger the insect to molt. If JH is present at the time of molting, the insect molts into a larger larval form; if absent, it molts into a pupa or adult. IGRs that mimic JH can produce premature molting of young immature stages, disrupting larval development. They can also act on eggs, causing sterility, disrupting behavior or disrupting diapause, the process that causes an insect to become dormant before winter. IGRs that inhibit JH production can cause insects to prematurely molt into a nonfunctional adult. IGRs that inhibit ecdysone can cause pupal mortality by interrupting the transformation of larval tissues into adult tissues during the pupal stage.

Chitin Synthesis Inhibitors

Chitin synthesis inhibitors work by preventing the formation of chitin, a carbohydrate needed to form the insect's exoskeleton. With these inhibitors, an insect grows normally until it molts. The inhibitors prevent the new exoskeleton from forming properly, causing the insect to die. Death may be quick, or take up to several days depending on the insect. Chitin synthesis inhibitors can also kill eggs by disrupting normal embryonic development. Chitin synthesis inhibitors affect insects for longer periods of time than hormonal IGRs. These are also quicker acting but can affect predaceous insects, arthropods and even fish. Compounds include benzoylurea pesticides.

Antimicrobial

An antimicrobial is an agent that kills microorganisms or inhibits their growth. Antimicrobial medicines can be grouped according to the microorganisms they act primarily against. For example, antibiotics are used against bacteria and antifungals are used against fungi. They can also be classified according to their function. Agents that kill microbes are called microbicidal, while those that merely inhibit their growth are called biostatic. The use of antimicrobial medicines to treat infection is known as antimicrobial chemotherapy, while the use of antimicrobial medicines to prevent infection is known as antimicrobial prophylaxis.

The main classes of antimicrobial agents are disinfectants ("nonselective antimicrobials" such as bleach), which kill a wide range of microbes on non-living surfaces to prevent the spread of illness, antiseptics (which are applied to living tissue and help reduce infection during surgery), and antibiotics (which destroy microorganisms within the body). The term "antibiotic" originally described only those formulations derived from living organisms but is now also applied to synthetic antimicrobials, such as the sulphonamides, or fluoroquinolones. The term also used to be restricted to antibacterials (and is often used as a synonym for them by medical professionals and in medical literature), but its context has broadened to include all antimicrobials. Antibacterial agents can be further subdivided into bactericidal agents, which kill bacteria, and bacteriostatic agents, which slow down or stall bacterial growth.

Use of substances with antimicrobial properties is known to have been common practice for at least 2000 years. Ancient Egyptians and ancient Greeks used specific molds and plant extracts to treat infection. More recently, microbiologists such as Louis Pasteur and Jules Francois Joubert observed antagonism between some bacteria and discussed the merits of controlling these interactions in medicine. In 1928, Alexander Fleming became the first to discover a natural antimicrobial fungus known as *Penicillium rubens*. The substance extracted from the fungus he named penicillin and in 1942 it was successfully used to treat a *Streptococcus* infection. Penicillin also proved successful in the treatment of many other infectious diseases such as gonorrhea, strep throat and pneumonia, which were potentially fatal to patients until then.

Many antimicrobial agents exist, for use against a wide range of infectious diseases.

Chemical

Selman Waksman, who was awarded the Nobel Prize in Medicine for developing 22 antibiotics—most notably Streptomycin.

Antibacterials

Antibacterials are used to treat bacterial infections. The toxicity to humans and other animals from antibacterials is generally considered low. However, prolonged use of certain antibacterials can decrease the number of gut flora, which may have a negative impact on health. After prolonged antibacterial use consumption of probiotics and reasonable eating can help to replace destroyed gut flora. Stool transplants may be considered for patients who are having difficulty recovering from prolonged antibiotic treatment, as for recurrent *Clostridium difficile* infections.

The discovery, development and clinical use of antibacterials during the 20th century has substantially reduced mortality from bacterial infections. The antibiotic era began with the pneumatic application of nitroglycerine drugs, followed by a "golden" period of discovery from about 1945 to 1970, when a number of structurally diverse and highly effective agents were discovered and developed. However, since 1980 the introduction of new antimicrobial agents for clinical use has declined, in part because of the enormous expense of developing and testing new drugs. Paralleled to this there has been an alarming increase in resistance of bacteria, fungi, viruses and parasites to multiple existing agents.

Antibacterials are among the most commonly used drugs; however antibiotics are also among the drugs commonly misused by physicians, such as usage of antibiotic agents in viral respiratory tract infections. As a consequence of widespread and injudicious use of antibacterials, there has been an accelerated emergence of antibiotic-resistant pathogens, resulting in a serious threat to global public health. The resistance problem demands that a renewed effort be made to seek antibacterial agents effective against pathogenic bacteria resistant to current antibacterials. Possible strategies towards this objective include increased sampling from diverse environments and application of metagenomics to identify bioactive compounds produced by currently unknown and uncultured microorganisms as well as the development of small-molecule libraries customized for bacterial targets.

Antifungals

Antifungals are used to kill or prevent further growth of fungi. In medicine, they are used as a treatment for infections such as athlete's foot, ringworm and thrush and work by exploiting differences between mammalian and fungal cells. They kill off the fungal organism without dangerous effects on the host. Unlike bacteria, both fungi and humans are eukaryotes. Thus, fungal and human cells are similar at the molecular level, making it more difficult to find a target for an antifungal drug to attack that does not also exist in the infected organism. Consequently, there are often side effects to some of these drugs. Some of these side effects can be life-threatening if the drug is not used properly.

As well as their use in medicine, antifungals are frequently sought after to control mold growth in damp or wet home materials. Sodium bicarbonate (baking soda) blasted on to surfaces acts as an antifungal. Another antifungal serum applied after or without blasting by soda is a mix of hydrogen peroxide and a thin surface coating that neutral-izes mold and encapsulates the surface to prevent spore release. Some paints are also manufactured with an added antifungal agent for use in high humidity areas such as bathrooms or kitchens. Other antifungal surface treatments typically contain variants of metals known to suppress mold growth e.g. pigments or solutions containing copper, silver or zinc. These solutions are not usually available to the general public because of their toxicity.

Antivirals

Antiviral drugs are a class of medication used specifically for treating viral infections. Like antibiotics, specific antivirals are used for specific viruses. They are relatively harmless to the host and therefore can be used to treat infections. They should be dis-tinguished from viricides, which actively deactivate virus particles outside the body.

Many of the antiviral drugs available are designed to treat infections by retroviruses, mostly HIV. Important antiretroviral drugs include the class of protease inhibitors. Herpes viruses, best known for causing cold sores and genital herpes, are usually treat-ed with the nucleoside analogue acyclovir. Viral hepatitis (A-E) are caused by five un-related hepatotropic viruses and are also commonly treated with antiviral drugs de-pending on the type of infection. influenza A and B viruses are important targets for the development of new influenza treatments to overcome the resistance to existing neuraminidase inhibitors such as oseltamivir.

Antiparasitics

Antiparasitics are a class of medications indicated for the treatment of infection by parasites, such as nematodes, cestodes, trematodes, infectious protozoa, and amoebae. Like antifungals, they must kill the infecting pest without serious damage to the host.

Non-Pharmaceutical

A wide range of chemical and natural compounds are used as antimicrobials. Organic acids are used widely as antimicrobials in food products, e.g. lactic acid, citric acid, acetic acid, and their salts, either as ingredients, or as disinfectants. For example, beef carcasses often are sprayed with acids, and then rinsed or steamed, to reduce the prevalence of $E.\ coli$.

Copper-alloy surfaces have natural intrinsic antimicrobial properties and can kill mi-croorganisms such as $E.\ coli$, MRSA and $Staphylococcus$. The United States Environ-mental Protection Agency has approved the registration of 355 such antibacterial cop-per alloys. As a public hygienic measure in addition to regular cleaning, antimicrobial copper alloys are being installed in healthcare facilities and in subway transit systems. Other heavy metal cations such as Hg^{2+} and Pb^{2+} have antimicrobial activities, but can be toxic to other living organisms such as humans.

Traditional herbalists used plants to treat infectious disease. Many of these plants have been investigated scientifically for antimicrobial activity, and some plant products have been shown to inhibit the growth of pathogenic microorganisms. A number of these agents appear to have structures and modes of action that are distinct from those of the antibiotics in current use, suggesting that cross-resistance with agents already in use may be minimal.

Essential Oils

Many essential oils included in herbal pharmacopoeias are claimed to possess antimicrobial activity, with the oils of bay, cinnamon, clove and thyme reported to be the most potent in studies with foodborne bacterial pathogens. Active constituents include terpenoid chemicals and other secondary metabolites. Despite their prevalent use in alternative medicine, essential oils have seen limited use in mainstream medicine. While 25 to 50% of pharmaceutical compounds are plant-derived, none are used as antimicrobials, though there has been increased research in this direction. Barriers to increased usage in mainstream medicine include poor regulatory oversight and quality control, mislabeled or misidentified products, and limited modes of delivery.

Antimicrobial Pesticides

According to the U.S. Environmental Protection Agency (EPA), and defined by the Federal Insecticide, Fungicide, and Rodenticide Act, antimicrobial pesticides are used in order to control growth of microbes through disinfection, sanitation, or reduction of development and to protect inanimate objects, industrial processes or systems, surfaces, water, or other chemical substances from contamination, fouling, or deterioration caused by bacteria, viruses, fungi, protozoa, algae, or slime.

Antimicrobial pesticide products The EPA monitors products, such as disinfectants/sanitizers for use in hospitals or homes, in order to ascertain efficacy. Products that are meant for public health are therefore under this monitoring system—ones used for drinking water, swimming pools, food sanitation, and other environmental surfaces. These pesticide products are registered under the premise that, when used properly, they do not demonstrate unreasonable side effects to humans or the environment. Even once certain products are on the market, the EPA continues to monitor and evaluate them to make sure they maintain efficacy in protecting public health.

Public health products regulated by the EPA are divided into three categories:

- Sterilizers (Sporicides): Will eliminate all bacteria, fungi, spores, and viruses.

- Disinfectants: Destroy or inactivate microorganisms (bacteria, fungi, viruses,) but may not act as sporicides (as those are the most difficult form to destroy). According to efficacy data, the EPA will classify a disinfectant as limited, general/broad spectrum, or as a hospital disinfectant.

- Sanitizers: Reduce the number of microorganisms, but may not kill or eliminate all of them.

Antimicrobial pesticide safety According to a 2010 CDC report, health-care workers can take steps to improve their safety measures against antimicrobial pesticide exposure. Workers are advised to minimize exposure to these agents by wearing protective equipment, gloves, and safety glasses. Additionally, it is important to follow the handling instructions properly, as that is how the Environmental Protection Agency has deemed it as safe to use. Employees should be educated about the health hazards, and encouraged to seek medical care if exposure occurs.

Ozone

Ozone can kill microorganisms in air and water, such as municipal drinking-water systems, swimming pools and spas, and the laundering of clothes.

Physical

Heat

Both dry and moist heat are effective in eliminating microbial life. For example, jars used to store preserves such as jam can be sterilized by heating them in a conventional oven. Heat is also used in pasteurization, a method for slowing the spoilage of foods such as milk, cheese, juices, wines and vinegar. Such products are heated to a certain temperature for a set period of time, which greatly reduces the number of harmful microorganisms.

Radiation

Foods are often irradiated to kill harmful pathogens. Common sources of radiation used in food sterilization include cobalt-60 (a gamma emitter), electron beams and x-rays. Ultraviolet light is also used to disinfect drinking water, both in small scale personal-use systems and larger scale community water purification systems.

Insect Repellent

An insect repellent (also commonly called "bug spray") is a substance applied to skin, clothing, or other surfaces which discourages insects (and arthropods in general) from landing or climbing on that surface. Insect repellents help prevent and control the outbreak of insect-borne (and other arthropod-bourne) diseases such as malaria, Lyme disease, dengue fever, bubonic plague, and West Nile fever. Pest animals commonly serving as vectors for disease include insects such as flea, fly, and mosquito; and the arachnid tick.

Common Insect Repellents

Oil Jar in cow horn for mosquito-repelling pitch oil, a by-product of the distillation of wood tar. Carried in a leather strap on a belt. Råneå, Norrbotten, since 1921 in Nordiska museet, Stockholm.

- Birch tree bark is traditionally made into tar. Combined with another oil (e.g., fish oil) at 1/2 dilution, it is then applied to the skin for repelling mosquitos

- DEET (*N,N*-diethyl-*m*-toluamide)

- Essential oil of the lemon eucalyptus (*Corymbia citriodora*) and its active compound p-menthane-3,8-diol (PMD)

- Icaridin, also known as picaridin, Bayrepel, and KBR 3023

- Nepetalactone, also known as "catnip oil"

- Citronella oil

- Neem oil

- Bog Myrtle (Myrica Gale)

- Dimethyl carbate

- Tricyclodecenyl allyl ether, a compound often found in synthetic perfumes.

- IR3535 (3-[N-Butyl-N-acetyl]-aminopropionic acid, ethyl ester)

- Ethylhexanediol, also known as Rutgers 612 or "6-12 repellent," discontinued in the US in 1991 due to evidence of causing developmental defects in animals

- Dimethyl phthalate, not as common as it once was but still occasionally an active ingredient in commercial insect repellents

- Metofluthrin

- Indalone. Widely used in a "6-2-2" mixture (60% Dimethyl phthalate, 20% Indalone, 20% Ethylhexanediol) during the 1940s and 1950s before the commercial introduction of DEET.

- Permethrin is different in that it is actually a contact insecticide.

- A more recent repellent being currently researched is SS220, which has been shown to provide significantly better protection than DEET.

- Another new and promising group of repellents are the anthranilate-based insect repellents.

Repellent Effectiveness

Synthetic repellents tend to be more effective and/or longer lasting than "natural" repellents. In comparative studies, IR3535 was as effective or better than DEET in protection against mosquitoes. Other sources (official publications of the associations of German physicians as well as of German druggists suggest the contrary and state DEET is still the most efficient substance available and the substance of choice for stays in malaria regions, while IR3535 has little effect. However, some plant-based repellents may provide effective relief as well. Essential oil repellents can be short-lived in their effectiveness, since essential oils can evaporate completely.

A popular post-WWII Australian brand of insect repellent.

A test of various insect repellents by an independent consumer organization found that repellents containing DEET or picaridin are more effective than repellents with "natural" active ingredients. All the synthetics gave almost 100% repellency for the first 2 hours, where the natural repellent products were most effective for the first 30 to 60 minutes, and required reapplication to be effective over several hours.

For protection against mosquitos, the U.S. Centers for Disease Control (CDC) issued a statement in May 2008 recommending equally DEET, picaridin, oil of lemon eucalyptus and IR3535 for skin. Permethrin is recommended for clothing, gear, or bed nets. In an earlier report, the CDC found oil of lemon eucalyptus to be more effective than other plant-based treatments, with a similar effectiveness to low concentrations of DEET. However, a 2006 published study found in both cage and field studies that a product containing 40% oil of lemon eucalyptus was just as effective as products containing high concentrations of DEET. Research has also found that neem oil is mosquito repellent for up to 12 hours. Citronella oil's mosquito repellency has also been verified by research, including effectiveness in repelling *Aedes aegypti*, but requires reapplication after 30 to 60 minutes.

More recently, in 2015, Researchers at New Mexico State University tested 10 commercially available products for their effectiveness at repelling mosquitoes. On the mosquito *Aedes aegypti*, the vector of Zika virus, only one repellent that did not contain

DEET had a strong effect for the duration of the 240 minutes test: a lemon eucalyptus oil repellent. All DEET-containing mosquito repellents were active.

There are also products available based on sound production, particularly ultrasound (inaudibly high frequency sounds) which purport to be insect repellents. However, these electronic devices have been shown to be ineffective based on studies done by the EPA and many universities.

Repellent Safety

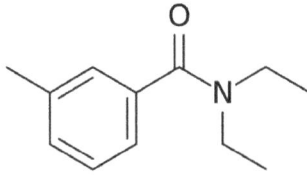

DEET

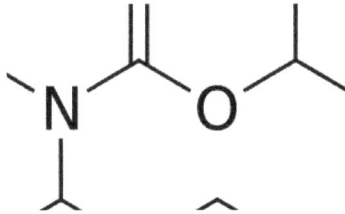

Icaridin

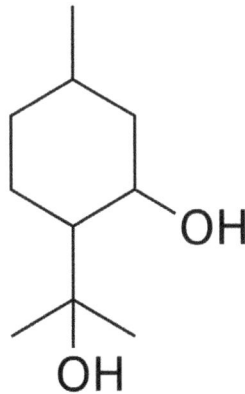

p-Menthane-3,8-diol (PMD)

Regarding safety with insect repellent use on children and pregnant women:

- Children may be at greater risk for adverse reactions to repellents, in part, because their exposure may be greater.

- Keep repellents out of the reach of children.

- Do not allow children to apply repellents to themselves.

- Use only small amounts of repellent on children.

- Do not apply repellents to the hands of young children because this may result in accidental eye contact or ingestion.

- Try to reduce the use of repellents by dressing children in long sleeves and long trousers tucked into boots or socks whenever possible. Use netting over strollers, playpens, etc.

- As with chemical exposures in general, pregnant women should take care to avoid exposures to repellents when practical, as the fetus may be vulnerable.

Some experts also recommend against applying chemicals such as DEET and sunscreen simultaneously since that would increase DEET penetration. Canadian researcher, Xiaochen Gu, a professor at the University of Manitoba's faculty of Pharmacy who led a study about mosquitos, advises that DEET should be applied 30 or more minutes later. Gu also recommends insect repellent sprays instead of lotions which are rubbed into the skin "forcing molecules into the skin".

Regardless of which repellent product used, it is recommended to read the label before use and carefully follow directions. Usage instructions for repellents vary from country to country. Some insect repellents are not recommended for use on younger children.

In the DEET Reregistration Eligibility Decision (RED) the United States Environmental Protection Agency (EPA) reported 14 to 46 cases of potential DEET associated seizures, including 4 deaths. The EPA states: "... it does appear that some cases are likely related to DEET toxicity," but observed that with 30% of the US population using DEET, the likely seizure rate is only about one per 100 million users.

The Pesticide Information Project of Cooperative Extension Offices of Cornell University states that, "Everglades National Park employees having extensive DEET exposure were more likely to have insomnia, mood disturbances and impaired cognitive function than were lesser exposed co-workers".

The EPA states that citronella oil shows little or no toxicity and has been used as a topical insect repellent for 60 years. However, the EPA also states that citronella may irritate skin and cause dermatitis in certain individuals. Canadian regulatory authorities concern with citronella based repellents is primarily based on data-gaps in toxicology, not on incidents.

Within countries of the European Union, implementation of Regulation 98/8/EC, commonly referred to as the Biocidal Products Directive, has severely limited the number and type of insect repellents available to European consumers. Only a small number of active ingredients have been supported by manufacturers in submitting dossiers to the EU Authorities.

In general, only formulations containing DEET, icaridin (sold under the trade name Saltidin and formerly known as Bayrepel or KBR3023), IR3535 (3-[N-Butyl-N-acetyl]-aminopropionic acid, ethyl ester) and Citriodiol (p-menthane-3,8-diol) are available. Most "natural" insect repellents such as citronella, neem oil, and herbal extracts are no longer permitted for sale as insect repellents in the EU; this does not preclude them from being sold for other purposes, as long as the label does not indicate they are a biocide (insect repellent).

Insect Repellents from Natural Sources

Mosquito repellent made from plants.

There are many preparations from naturally occurring sources that have been used as a repellent to certain insects. Some of these act as insecticides while others are only repellent.

- *Achillea alpina* (mosquitos)
- alpha-terpinene (mosquitos)
- Basil
- Sweet Basil (*Ocimum basilicum*)
- *Callicarpa americana* (Beautyberry)
- Breadfruit (Insect repellent, including mosquitoes)
- Camphor (moths)
- Carvacrol (mosquitos)
- Castor oil (*Ricinus communis*) (mosquitos)
- Catnip oil (*Nepeta* species) (nepetalactone against mosquitos) (WARNING: may attract cats)
- Cedar oil (mosquitos, moths)
- Celery extract (*Apium graveolens*) (mosquitos) In clinical testing an extract of

celery was demonstrated to be at least equally effective to 25% DEET, although the commercial availability of such an extract is not known.

- Cinnamon (leaf oil kills mosquito larvae)

- Citronella oil (repels mosquitos)

- Oil of cloves (mosquitos)

- Eucalyptus oil (70%+ eucalyptol), (cineol is a synonym), mosquitos, flies, dust mites)

- Fennel oil (*Foeniculum vulgare*) (mosquitos)

- Garlic (*Allium sativum*) (Mosquito, rice weevil, wheat flour beetle)

- Geranium oil (also known as *Pelargonium graveolens*)

- Lavender (ineffective alone, but measurable effect in certain repellent mixtures)

- Lemon eucalyptus (*Corymbia citriodora*) essential oil and its active ingredient p-menthane-3,8-diol (PMD)

- Lemongrass oil (*Cymbopogon* species) (mosquitos)

- East-Indian Lemon Grass (*Cymbopogon flexuosus*)

- Marigolds (*Tagetes* species)

- Marjoram (Spider mites *Tetranychus urticae* and *Eutetranychus orientalis*)

- Neem oil (*Azadirachta indica*) (Repels or kills mosquitos, their larvae and a plethora of other insects including those in agriculture)

- Oleic acid, repels bees and ants by simulating the "Smell of death" produced by their decomposing corpses.

- Pennyroyal (*Mentha pulegium*) (mosquitos, fleas), but very toxic to pets.

- Peppermint (*Mentha* x *piperita*) (mosquitos)

- Pyrethrum (from *Chrysanthemum* species, particularly *C. cinerariifolium* and *C. coccineum*)

- Rosemary (*Rosmarinus officinalis*) (mosquitos)

- Spanish Flag (*Lantana camara*) (against Tea Mosquito Bug, *Helopeltis theivora*)

- Tea tree oil from the leaves of *Melaleuca alternifolia*

- Thyme (*Thymus* species) (mosquitos)

- Yellow Nightshade (*Solanum villosum*), berry juice (against *Stegomyia aegypti* mosquitos)

- *Andrographis paniculata* extracts (mosquito)

Inactive Substances – Carriers

In 2002, the *New England Journal of Medicine* published an article that found products containing essential oils such as catnip or geranium oil, when combined with a suitable carrier oil such as soybean, have been found to be effective as natural repellents. This was based on testing done by Johns Hopkins and Cornell Universities. Other commercial products offered for household mosquito "control" include small electrical mats, mosquito repellent vapor, DEET-impregnated wrist bands, mosquito fogging, and mosquito coils containing a form of the chemical allethrin. Mosquito-repellent candles containing citronella oil are sold widely in the U.S. These have been used with mixed reports of success and failure.

Less Effective Methods

Some old studies suggested that the ingestion of large doses of thiamine could be effective as an oral insect repellent against mosquito bites. However, there is now conclusive evidence that thiamin has no efficacy against mosquito bites. Some claim that plants like wormwood or sagewort, lemon balm, lemon grass, lemon thyme and the mosquito plant (Pelargonium) will act against mosquitoes. However, scientists have determined that these plants are "effective" for a limited time only when the leaves are crushed and applied directly to the skin.

There are several, widespread, unproven theories about mosquito control, such as the assertion that vitamin B, in particular B_1 (thiamine), garlic, ultrasonic devices or incense can be used to repel or control mosquitoes. Moreover, manufacturers of "mosquito repelling" ultrasonic devices have been found to be fraudulent, and their devices were deemed "useless" in tests by the UK Consumer magazine *Which?*, and according to a review of scientific studies.

Animal Repellent

Animal repellents are products designed to keep certain animals away from objects, areas, people, plants, or other animals.

Scarecrow in a field

Tat guards, steel or aluminum discs, attached to the mooring line to prevent rats from boarding a ship

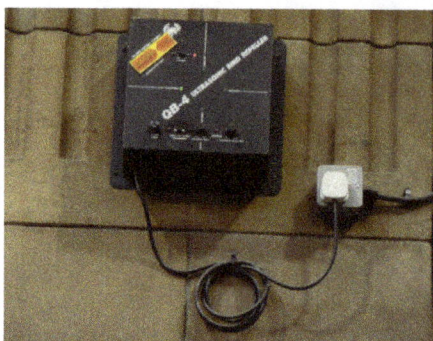

Ultrasonic bird repeller

Overview

Repellents generally work by taking advantage of an animal's natural aversion to something, and often the thing chosen is something that the animal has learned to avoid (or instinctively avoids) in its natural environment.

For example, some animals will avoid anything that has the odor of the urine of certain predators. Tiger urine is thus very effective at keeping away animals. Coyote urine has gained currency as a deer repellent. Fox urine is used to repel rabbits, groundhogs, woodchucks, squirrels and chipmunks. Bobcat urine repels moles, mice, voles and other rodents. Wolf urine is used to repel moose.

Chemical repellents mimic natural substances that repel or deter animals, or they are designed to be so irritating to a specific animal or type of animal that the targeted animal will avoid the protected object or area. Some chemical repellents combine both principles. There are many homemade deer repellent recipes on the web.

For example, the lawn fertilizer Milorganite is claimed to be an effective repellent due to its smell. Repellents fall into two main categories, odor and taste. Odor repellents work better in the warm seasons and taste repellents work better in the cold months. Taste repellents only work after the deer or other animal has taken a bit out of the plant. If you have a plant you don't want nibbled on at all, use an odor repellent or a combination of both.

Other types of non-chemical repellents are sometimes used. A simple electrified or barbed wire fence can mechanically repel livestock or predator animals. Some electrical repellent systems have been tested against sharks. High-frequency whistles have been used on vehicles to drive deer away from highways, and similar devices have been used to deter and repel certain types of insects or rodents. Repellents for domestic cats and dogs can also be found; these include ultrasonic devices which emit a high frequency noise that does not affect humans. These types of non-chemical repellents are quite controversial because their effectiveness varies from person to person. Furthermore, there have been few scientific studies conducted to prove that they do work. They are, however, a safe and humane way of disposing pests. Flashing lights are used to repel lions in Kenya.

The ideal repellent is completely specific for the target animal; that is, it drives away the animal that one wishes to repel *without* affecting or harming any other animals or people. One type of animal repellent may be effective for raccoons, while another animal repellent may be more effective for skunks. It can be difficult to design a repellent method that drives away only undesirable animals while having no effect on people or other creatures.

Some animals are more likely to be targeted than others by human users of repellents. Targeted animals are often predators of animals of interest to human beings, such as food fish or livestock. Sometimes the targeted animals are predators of human beings themselves.

Snake Repellents

- Research has shown that cinnamon oil, clove oil, and eugenol are effective snake repellents. Snakes will retreat when sprayed directly with these oils and will exit cargo or other confined spaces when these oils are introduced to the area.

- In ancient times the Greek historian Herodotus noted that Arabian harvesters of frankincense used burning resin from *Styrax* trees to repel poisonous snakes that lived in the trees.

- Camphor

- Moth balls

- The roots and other parts of *Acacia polyacantha subsp. campylacantha* emit chemical compounds that repel animals including rats, snakes and crocodiles. For snakes, roots are placed in the rafters of houses.

References

- Francis Borgio J, Sahayaraj K and Alper Susurluk I (eds) . Microbial Insecticides: Principles and Applications, Nova Publishers, USA. 492pp. ISBN 978-1-61209-223-2.

• Metcalf, Robert L. (2002). "Ullmann's Encyclopedia of Industrial Chemistry". Ullmann's Encyclopedia of Industrial Chemistry. Wiley-VCH. doi:10.1002/14356007.a14_263. ISBN 3527306730.

• Kramer, Wolfgang and Schirmer, Ulrich (2012) Modern Crop Protection Compounds. Wiley pp. 197-276. ISBN 978-3-527-32965-6

• Powles, S. B.; Shaner, D. L., eds. (2001). Herbicide Resistance and World Grains. CRC Press, Boca Raton, FL. p. 328. ISBN 9781420039085.

• Moss, S. R. (2002). "Herbicide-Resistant Weeds". In Naylor,, R. E. L. Weed management handbook (9th ed.). Blackwell Science Ltd. pp. 225–252. ISBN 0-632-05732-7.

• Sandle T (editor) (2012). The CDC Handbook: A Guide to Cleaning and Disinfecting Cleanrooms (1st ed.). Grosvenor House Publishing Limited. ISBN 978-1781487686.

• "Common Cleaning Products May Be Dangerous When Mixed" (PDF). New Jersey Department of Health and Senior Services. Retrieved 19 April 2016.

• "Division of Oral Health - Infection Control Glossary". U.S. Centers for Disease Control and Prevention. Retrieved 19 April 2016

• "Island which spent £600,000 getting rid of rats over-run with rabbits". Daily Telegraph. 27 April 2010. Retrieved 4 April 2015.

• "CDC - Pesticide Illness & Injury Surveillance - NIOSH Workplace Safety and Health Topic". Cdc. gov. 2013-09-11. Retrieved 2014-01-28.

• "Pesticides 101 - A primer on pesticides, their use in agriculture and the exposure we face | Pesticide Action Network". Panna.org. Retrieved 2014-01-28.

• "EU approves powdered corn cob as biocidal active". Chemical Watch: Global Risk & Regulation News. 15 August 2013. Retrieved 22 August 2013.

Plant Disease Resistance and Defense

Plant disease resistance is the reduction of pathogen growth on or in the plant, at the same time plant disease defense is a range of adaptions evolved by plants, which improve their survival. The aspects elucidated in this section are of vital importance, and provide a better understanding on the subject matter.

Plant Disease Resistance

Plant disease resistance protects plants from pathogens in two ways: mechanisms and by infection-induced responses of the immune system. Relative to a susceptible plant, disease resistance is the reduction of pathogen growth on or in the plant, while the term disease tolerance describes plants that exhibit little disease damage despite substantial pathogen levels. Disease outcome is determined by the three-way interaction of the pathogen, the plant and the environmental conditions (an interaction known as the disease triangle).

Defense-activating compounds can move cell-to-cell and systemically through the plant vascular system. However, plants do not have circulating immune cells, so most cell types exhibit a broad suite of antimicrobial defenses. Although obvious *qualitative* differences in disease resistance can be observed when multiple specimens are compared (allowing classification as "resistant" or "susceptible" after infection by the same pathogen strain at similar inoculum levels in similar environments), a gradation of *quantitative* differences in disease resistance is more typically observed between plant strains or genotypes. Plants consistently resist certain pathogens but succumb to others; resistance is usually pathogen species- or pathogen strain-specific.

Background

Plant disease resistance is crucial to the reliable production of food, and it provides significant reductions in agricultural use of land, water, fuel and other inputs. Plants in both natural and cultivated populations carry inherent disease resistance, but this has not always protected them.

The late blight Irish potato famine of the 1840s was caused by the oomycete Phytophthora infestans. The world's first mass-cultivated banana cultivar Gros Michel was lost in the 1920s to Panama disease caused by the fungus Fusarium oxysporum. The current wheat stem, leaf, and yellow stripe rust epidemics spreading from East Africa into the Indian subcontinent are caused by rust fungi Puccinia graminis and P. striiformis. Other epidemics include Chestnut blight, as well as recurrent severe plant diseases such as Rice blast, Soybean cyst nematode, Citrus canker.

Plant pathogens can spread rapidly over great distances, vectored by water, wind, insects, and humans. Across large regions and many crop species, it is estimated that diseases typically reduce plant yields by 10% every year in more developed nations or agricultural systems, but yield loss to diseases often exceeds 20% in less developed settings, an estimated 15% of global crop production.

However, disease control is reasonably successful for most crops. Disease control is achieved by use of plants that have been bred for good resistance to many diseases, and by plant cultivation approaches such as crop rotation, pathogen-free seed, appropriate planting date and plant density, control of field moisture and pesticide use.

Viral Disease Common Mechanisms

Pre-formed structures and compounds

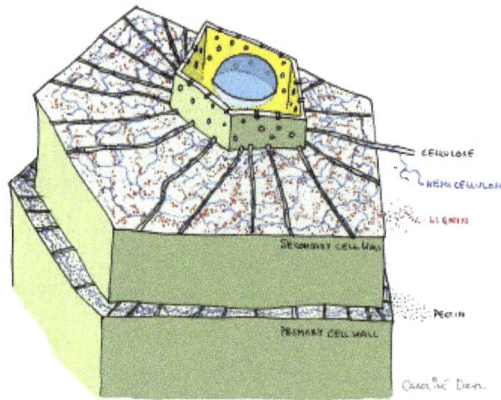

secondary plant wall

- Plant cuticle/surface

- Plant cell walls

- Antimicrobial chemicals (for example: glucosides, saponins)

- Antimicrobial proteins

- Enzyme inhibitors

- Detoxifying enzymes that break down pathogen-derived toxins

- Receptors that perceive pathogen presence and activate inducible plant defences

Inducible Post-Infection Plant Defenses

- Cell wall reinforcement (callose, lignin, suberin, cell wall proteins)

- Antimicrobial chemicals, including reactive oxygen species such as hydrogen peroxide or peroxynitrite, or more complex phytoalexins such as genistein or camalexin

- Antimicrobial proteins such as defensins, thionins, or PR-1

- Antimicrobial enzymes such as chitinases, beta-glucanases, or peroxidases

- Hypersensitive response - a rapid host cell death response associated with defence mediated by "Resistance genes."(Bryant, Tracy 2008).

Variable Resistance

Even in susceptible plants to obligate parasites the tissue resistance changes due to ontogeny and to influence of external conditions. This resistance can be measured by the value of redox potential of electron carriers, which is produced in the plant by enzymatic reactions associated with respiration. Electron carriers are water-soluble and are not oxidized by air oxygen. There is not free oxygen in the cells, all the oxidations and reductions take place enzymatically. These reactions are highly specific for the plant species. The host and parasite have different electron carriers.

Immune System

The plant immune system consists of two interconnected tiers of receptors, one outside and one inside the cell. Both systems sense the intruder, respond to the intrusion and optionally signal to the rest of the plant and sometimes to neighboring plants that the intruder is present. The two systems detect different types of pathogen molecules and classes of plant receptor proteins

The first tier is primarily governed by pattern recognition receptors that are activated by recognition of evolutionarily conserved pathogen or microbial–associated molecular patterns (PAMPs or MAMPs, here P/MAMP). Activation of PRRs leads to intracellular signaling, transcriptional reprogramming, and biosynthesis of a complex output response that limits colonization. The system is known as PAMP-Triggered Immunity (PTI)"(JonesDangl2010).

The second tier (again, primarily), effector-triggered immunity (ETI), consists of another set of receptors, (nucleotide-binding)They operate within the cell, encoded by R genes. The presence of specific pathogen "effectors" activates specific NLR proteins that limit pathogen proliferation.

Receptor responses include ion channel gating, oxidative burst, cellular redox changes, or protein kinase cascades that directly activate cellular changes (such as cell wall reinforcement or antimicrobial production), or activate changes in gene expression that then elevate other defensive responses

Plant immune systems show some mechanistic similarities with the immune systems of insects and mammals, but also exhibit many plant-specific characteristics. Plants can sense the presence of pathogens and the effects of infection via activated by touch]. Rice Universi

PAMp-Triggered Immunity

PAMP-Triggered Immunity conserved molecules that inhabit multiple pathogen genera are classified as MAMPs by some researchers. The defenses induced by MAMP perception are sufficient to repel most pathogens. However, pathogen effector proteins are adapted to suppress basal defenses such as PTI

Effector Triggered Immunity

Effector Triggered Immunity (ETI) is activated by the presence of pathogen effectors. The ETI immune response is reliant on R genes, and is activated by specific pathogen strains. As with PTI, many specific examples of apparent ETI violate commoMost plant immune systems carry a repertoire of 100-600 different R genes that mediate resistance to various virus .Plant ETI often cause an apoptotic hypersensitive response. (Odds Rathjen2010).

R Genes and R Proteins

Plants have evolved R genes (resistance genes) whose products allow recognition of specific pathogen effectors, either through direct binding or by recognition of the effector's alteration of a host protein. These virulence factors drove co-evolution of plant resistant genes to combat the pathogens' Avr (avirulent) genes. Many R genes encode NB-LRR proteins (nucleotide-binding/leucine-rich repeat domains, also known as NLR proteins or STAND proteins, among other names).

R gene products control a broad set of disease resistance responses whose induction is often sufficient to stop further pathogen growth/spread. Each plant genome contains a few hundred apparent R genes. Studied R genes usually confer specificity for particular pathogen strains. As first noted by Harold Flor in his mid-20th century formulation of the gene-for-gene relationship, the plant R gene and the pathogen Avr gene must have matched specificity for that R gene to confer resistance, suggesting a receptor/ligand interaction for Avr and R genes. Alternatively, an effector can modify its host cellular target (or a molecular decoy of that target) activating an NLR associated with the

Effector Biology

So-called "core" effectors are defined operationally by their wide distribution across the population of a particular pathogen and their substantial contribution to pathogen virulence. Genomics can be used to identify core effectors, which can then functionally define new R alleles, which can serve as breeding targets.

RNA Silencing and Systemic Acquired Resistance Elicited by Prior Infections

Against viruses, plants often induce pathogen-specific gene silencing mechanisms mediated by RNA interference. T

Plant immune systems also can respond to an initial infection in one part of the plant by physiologically elevating the capacity for a successful defense response in other parts. Such responses include systemic acquired resistance, largely mediated by salicylic acid-dependent pathways, and induced systemic resistance, largely mediated.

Species-Level Resistance

In a small number of cases, plant genes are effective against an entire pathogen species, even though that species that is pathogenic on other genotypes of that host species. Examples include barley MLO against powdery mildew, wheat Lr34 against leaf rust and wheat Yr36 against stripe rust. An array of mechanisms for this type of resistance may exist depending on the particular gene and plant-pathogen combination. Other reasons for effective plant immunity can include a lack of coadaptation (the pathogen and/or plant lack multiple mechanisms needed for colonization and growth within that host species), or a particularly effective suite of pre-formed defenses.

Signaling Mechanisms

Perception of Pathogen Presence

Plant defense signaling is activated by pathogen-detecting receptors. The activated receptors frequently elicit reactive oxygen and nitric oxide production, calcium, potassium and proton ion fluxes, altered levels of salicylic acid and other hormones and activation of MAP kinases and other specific protein kinases. These events in turn typically lead to the modification of proteins that control gene transcription, and the activation of defense-associated gene expression.

In addition to PTI and ETI, plant defenses can be activated by the sensing of damage-associated compounds (DAMP), such as portions of the plant cell wall released during pathogenic infection. Many receptors for MAMPs, effectors and DAMPs have been discovered. Effectors are often detected by NLRs, while MAMPs and DAMPs are often detected by transmembrane receptor-kinases that carry LRR or LysM extracellular domains.

Transcription Factors and The Hormone Response

Numerous genes and/or proteins as well as other molecules have been identified that mediate plant defense signal transduction. Cytoskeleton and vesicle trafficking dynamics help to orient plant defense responses toward the point of pathogen attack.

Mechanisms of Transcription Factors and Hormones

Plant immune system activity is regulated in part by signaling hormones such as:

- Salicylic acid

- Jasmonic acid

- Ethylene

There can be substantial cross-talk among these pathways.

Regulation by Degradation

As with many signal transduction pathways, plant gene expression during immune responses can be regulated by degradation. This often occurs when hormone binding to hormone receptors stimulates ubiquitin-associated degradation of repressor proteins that block expression of certain genes. The net result is hormone-activated gene expression. Examples:

- Auxin: binds to receptors that then recruit and degrade repressors of transcriptional activators that stimulate auxin-specific gene expression.

- Jasmonic acid: similar to auxin, except with jasmonate receptors impacting jasmonate-response signaling mediators such as JAZ proteins.

- Gibberellic acid: Gibberellin causes receptor conformational changes and binding and degradation of Della proteins.

- Ethylene: Inhibitory phosphorylation of the EIN2 ethylene response activator is blocked by ethylene binding. When this phosphorylation is reduced, EIN2 protein is cleaved and a portion of the protein moves to the nucleus to activate ethylene-response gene expression.

Ubiquitin and E3 Signaling

Ubiquitination plays a central role in cell signaling that regulates processes including protein degradation and immunological response. Although one of the main functions of ubiquitin is to target proteins for destruction, it is also useful in signaling pathways, hormone release, apoptosis and translocation of materials throughout the cell. Ubiquitination is a component of several immune responses. Without ubiquitin's proper functioning, the invasion of pathogens and other harmful molecules would increase dramatically due to weakened immune defenses.

E3 Signaling

The E3 Ubiquitin ligase enzyme is a main component that provides specificity in protein degradation pathways, including immune signaling pathways. The E3 enzyme components can be grouped by which domains they contain and include several types. These include the Ring and U-box single subunit, HECT, and CRLs. Plant signaling pathways including immune responses are controlled by several feedback pathways, which often include negative feedback; and they can be regulated by De-ubiquitination enzymes, degradation of transcription factors and the degradation of negative regulators of transcription.

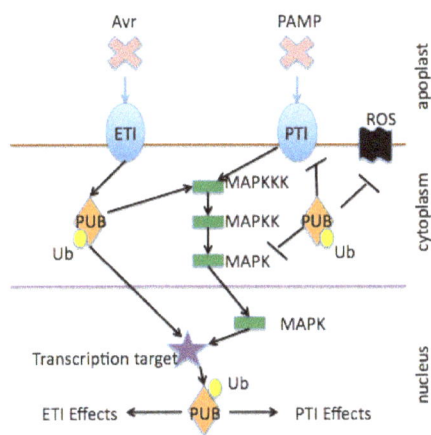

This image depicts the pathways taken during responses in plant immunity. It highlights the role and effect ubiquitin has in regulating the pathway.

Plant Breeding for Disease Resistance

Plant breeders emphasize selection and development of disease-resistant plant lines. Plant diseases can also be partially controlled by use of pesticides and by cultivation practices such as crop rotation, tillage, planting density, disease-free seeds and cleaning of equipment, but plant varieties with inherent (genetically determined) disease resistance are generally preferred. Breeding for disease resistance began when plants were first domesticated. Breeding efforts continue because pathogen populations are under selection pressure for increased virulence, new pathogens appear, evolving cultivation practices and changing climate can reduce resistance and/or strengthen pathogens, and plant breeding for other traits can disrupt prior resistance. A plant line with acceptable resistance against one pathogen may lack resistance against others.

Breeding for Resistance Typically Includes:

- Identification of plants that may be less desirable in other ways, but which carry a useful disease resistance trait, including wild strains that often express enhanced resistance.

- Crossing of a desirable but disease-susceptible variety to another variety that is a source of resistance.

- Growth of breeding candidates in a disease-conducive setting, possibly including pathogen inoculation. Attention must be paid to the specific pathogen isolates, to address variability within a single pathogen species.

- Selection of disease-resistant individuals that retain other desirable traits such as yield, quality and including other disease resistance traits.

Resistance is termed *durable* if it continues to be effective over multiple years of widespread use as pathogen populations evolve. "Vertical resistance" is specific to certain races or strains of a pathogen species, is often controlled by single R genes and can be less durable. Hoizontal or broad-spectrum resistance against an entire pathogen species is often only incompletely effective, but more durable, and is often controlled by many genes that segregate in breeding populations.

Crops such as potato, apple, banana and sugarcane are often propagated by vegetative reproduction to preserve highly desirable plant varieties, because for these species, outcrossing seriously disrupts the preferred traits. Vegetatively propagated crops may be among the best targets for resistance improvement by the biotechnology method of plant transformation to manage genes that affect disease resistance.

Scientific breeding for disease resistance originated with Sir Rowland Biffen, who identified a single recessive gene for resistance to wheat yellow rust. Nearly every crop was then bred to include disease resistance (R) genes, many by introgression from compatible wild relatives.

GM or Transgenic Engineered Disease Resistance

The term GM ("genetically modified") is often used as a synonym of transgenic to refer to plants modified using recombinant DNA technologies. Plants with transgenic/GM disease resistance against insect pests have been extremely successful as commercial products, especially in maize and cotton, and are planted annually on over 20 million hectares in over 20 countries worldwide. Transgenic plant disease resistance against microbial pathogens was first demonstrated in 1986. Expression of viral coat protein gene sequences conferred virus resistance via small RNAs. This proved to be a widely applicable mechanism for inhibiting viral replication. Combining coat protein genes from three different viruses, scientists developed squash hybrids with field-validated, multiviral resistance. Similar levels of resistance to this variety of viruses had not been achieved by conventional breeding.

A similar strategy was deployed to combat papaya ringspot virus, which by 1994 threatened to destroy Hawaii's papaya industry. Field trials demonstrated excellent efficacy and high fruit quality. By 1998 the first transgenic virus-resistant papaya was approved for sale. Disease resistance has been durable for over 15 years. Transgenic papaya accounts for ~85% of Hawaiian production. The fruit is approved for sale in the U.S., Canada and Japan.

Potato lines expressing viral replicase sequences that confer resistance to potato leafroll virus were sold under the trade names NewLeaf Y and NewLeaf Plus, and were widely accepted in commercial production in 1999-2001, until McDonald's Corp. decided not to purchase GM potatoes and Monsanto decided to close their NatureMark potato business. NewLeaf Y and NewLeaf Plus potatoes carried two GM traits, as they also expressed Bt-mediated resistance to Colorado potato beetle.

No other crop with engineered disease resistance against microbial pathogens had reached the market by 2013, although more than a dozen were in some state of development and testing.

Examples of transgenic disease resistance projects				
Publication year	Crop	Disease resistance	Mechanism	Development status
2012	Tomato	Bacterial spot	R gene from pepper	8 years of field trials
2012	Rice	Bacterial blight and bacterial streak	Engineered E gene	Laboratory
2012	Wheat	Powdery mildew	Overexpressed R gene from wheat	2 years of field trials at time of publication
2011	Apple	Apple scab fungus	Thionin gene from barley	4 years of field trials at time of publication
2011	Potato	Potato virus Y	Pathogen-derived resistance	1 year of field trial at time of publication
2010	Apple	Fire blight	Antibacterial protein from moth	12 years of field trials at time of publication
2010	Tomato	Multibacterial resistance	PRR from *Arabidopsis*	Laboratory scale
2010	Banana	Xanthomonas wilt	Novel gene from pepper	Now in field trial
2009	Potato	Late blight	R genes from wild relatives	3 years of field trials
2009	Potato	Late blight	R gene from wild relative	2 years of field trials at time of publication
2008	Potato	Late blight	R gene from wild relative	2 years of field trials at time of publication

2008	Plum	Plum pox virus	Pathogen-derived resistance	Regulatory approvals, no commercial sales
2005	Rice	Bacterial streak	R gene from maize	Laboratory
2002	Barley	Stem rust	Resting lymphocyte kinase (RLK) gene from resistant barley cultivar	Laboratory
1997	Papaya	Ring spot virus	Pathogen-derived resistance	Approved and commercially sold since 1998, sold into Japan since 2012
1995	Squash	Three mosaic viruses	Pathogen-derived resistance	Approved and commercially sold since 1994
1993	Potato	Potato virus X	Mammalian interferon-induced enzyme	3 years of field trials at time of publication

PRR Transfer

Research aimed at engineered resistance follows multiple strategies. One is to transfer useful PRRs into species that lack them. Identification of functional PRRs and their transfer to a recipient species that lacks an orthologous receptor could provide a general pathway to additional broadened PRR repertoires. For example, the *Arabidopsis* PRR *EF-Tu* receptor (EFR) recognizes the bacterial translation elongation factor *EF-Tu*. Research performed at Sainsbury Laboratory demonstrated that deployment of EFR into either *Nicotiana benthamiana* or *Solanum lycopersicum* (tomato), which cannot recognize *EF-Tu*, conferred resistance to a wide range of bacterial pathogens. EFR expression in tomato was especially effective against the widespread and devastating soil bacterium Ralstonia solanacearum. Conversely, the tomato PRR *Verticillium 1* (*Ve1*) gene can be transferred from tomato to *Arabidopsis*, where it confers resistance to race 1 Verticillium isolates.

Stacking

The second strategy attempts to deploy multiple NLR genes simultaneously, a breeding strategy known as stacking. Cultivars generated by either DNA-assisted molecular breeding or gene transfer will likely display more durable resistance, because pathogens would have to mutate multiple effector genes. DNA sequencing allows researchers to functionally "mine" NLR genes from multiple species/strains.

The *avrBs2* effector gene from *Xanthomona perforans* is the causal agent of bacterial spot disease of pepper and tomato. The first "effector-rationalized" search for a potentially durable R gene followed the finding that *avrBs2* is found in most disease-causing *Xanthomonas* species and is required for pathogen fitness. The *Bs2* NLR gene from the wild pepper, *Capsicum chacoense*, was moved into tomato, where it inhibited pathogen growth. Field trials demonstrated robust resistance without bactericidal chemicals. However, rare strains of *Xanthomonas* overcame *Bs2*-mediated resistance in pepper by acquisition of *avrBs2* mutations that avoid recognition but retain virulence. Stacking R genes that each recognize a different core effector could delay or prevent adaptation.

More than 50 loci in wheat strains confer disease resistance against wheat stem, leaf and yellow stripe rust pathogens. The Stem rust 35 (*Sr35*) NLR gene, cloned from a diploid relative of cultivated wheat, *Triticum monococcum*, provides resistance to wheat rust isolate *Ug99*. Similarly, *Sr33*, from the wheat relative *Aegilops tauschii*, encodes a wheat ortholog to barley *Mla* powdery mildew–resistance genes. Both genes are unusual in wheat and its relatives. Combined with the *Sr2* gene that acts additively with at least Sr33, they could provide durable disease resistance to *Ug99* and its derivatives.

Executor Genes

Another class of plant disease resistance genes opens a "trap door" that quickly kills invaded cells, stopping pathogen proliferation. Xanthomonas and Ralstonia transcription activator–like (TAL) effectors are DNA-binding proteins that activate host gene expression to enhance pathogen virulence. Both the rice and pepper lineages independently evolved TAL-effector binding sites that instead act as an executioner that induces hypersensitive host cell death when up-regulated. *Xa27* from rice and Bs3 and Bs4c from pepper, are such "executor" (or "executioner") genes that encode non-homologous plant proteins of unknown function. Executor genes are expressed only in the presence of a specific TAL effector.

Engineered executor genes were demonstrated by successfully redesigning the pepper *Bs3* promoter to contain two additional binding sites for TAL effectors from disparate pathogen strains. Subsequently, an engineered executor gene was deployed in rice by adding five TAL effector binding sites to the *Xa27* promoter. The synthetic *Xa27* construct conferred resistance against Xanthomonas bacterial blight and bacterial leaf streak species.

Host Susceptibility Alleles

Most plant pathogens reprogram host gene expression patterns to directly benefit the pathogen. Reprogrammed genes required for pathogen survival and proliferation can be thought of as "disease-susceptibility genes." Recessive resistance genes are disease-susceptibility candidates. For example, a mutation disabled an *Arabidopsis* gene encoding pectate lyase (involved in cell wall degradation), conferring resistance to the powdery mildew pathogen *Golovinomyces cichoracearum*. Similarly, the Barley *MLO* gene and spontaneously mutated pea and tomato *MLO* orthologs also confer powdery mildew resistance.

Lr34 is a gene that provides partial resistance to leaf and yellow rusts and powdery mildew in wheat. *Lr34* encodes an adenosine triphosphate (ATP)–binding cassette (ABC) transporter. The dominant allele that provides disease resistance was recently found in cultivated wheat (not in wild strains) and, like *MLO* provides broad-spectrum resistance in barley.

Natural alleles of host translation elongation initiation factors *eif4e* and *eif4g* are also recessive viral-resistance genes. Some have been deployed to control potyviruses in barley, rice, tomato, pepper, pea, lettuce and melon. The discovery prompted a successful mutant screen for chemically induced *eif4e* alleles in tomato.

Natural promoter variation can lead to the evolution of recessive disease-resistance alleles. For example, the recessive resistance gene *xa13* in rice is an allele of *Os-8N3*. *Os-8N3* is transcriptionally activated by*Xanthomonas oryzae pv. oryzae* strains that express the TAL effector *PthXo1*. The *xa13* gene has a mutated effector-binding element in its promoter that eliminates *PthXo1* binding and renders these lines resistant to strains that rely on *PthXo1*. This finding also demonstrated that *Os-8N3* is required for susceptibility.

Xa13/Os-8N3 is required for pollen development, showing that such mutant alleles can be problematic should the disease-susceptibility phenotype alter function in other processes. However, mutations in the *Os11N3* (OsSWEET14) TAL effector–binding element were made by fusing TAL effectors to nucleases (TALENs). Genome-edited rice plants with altered *Os11N3* binding sites remained resistant to *Xanthomonas oryzae pv. oryzae*, but still provided normal development function.

Gene Silencing

RNA silencing-based resistance is a powerful tool for engineering resistant crops. The advantage of RNAi as a novel gene therapy against fungal, viral and bacterial infection in plants lies in the fact that it regulates gene expression via messenger RNA degradation, translation repression and chromatin remodelling through small non-coding RNAs. Mechanistically, the silencing processes are guided by processing products of the double-stranded RNA (dsRNA) trigger, which are known as small interfering RNAs and microRNAs.

Host Range

Among the thousands of species of plant pathogenic microorganisms, only a small minority have the capacity to infect a broad range of plant species. Most pathogens instead exhibit a high degree of host-specificity. Non-host plant species are often said to express *non-host resistance*. The term *host resistance* is used when a pathogen species can be pathogenic on the host species but certain strains of that plant species resist certain strains of the pathogen species. The causes of host resistance and non-host resistance can overlap. Pathogen host range can change quite suddenly if, for example, the pathogen's capacity to synthesize a host-specific toxin or effector is gained by gene shuffling/mutation, or by horizontal gene transfer.

Epidemics and Population Biology

Native populations are often characterized by substantial genotype diversity and dispersed populations (growth in a mixture with many other plant species). They also have undergone of plant-pathogen coevolution. Hence as long as novel pathogens are not introduced/do not evolve, such populations generally exhibit only a low incidence of severe disease epidemics.

Monocrop agricultural systems provide an ideal environment for pathogen evolution, because they offer a high density of target specimens with similar/identical genotypes.

The rise in mobility stemming from modern transportation systems provides pathogens with access to more potential targets.

Climate change can alter the viable geographic range of pathogen species and cause some diseases to become a problem in areas where the disease was previously less important.

These factors make modern agriculture more prone to disease epidemics. Common solutions include constant breeding for disease resistance, use of pesticides, use of border inspections and plant import restrictions, maintenance of significant genetic diversity within the crop gene pool, and constant surveillance to accelerate initiation of appropriate responses. Some pathogen species have much greater capacity to overcome plant disease resistance than others, often because of their ability to evolve rapidly and to disperse broadly.

Plant Defense Against Herbivory

Plant defense against herbivory or host-plant resistance (HPR) describes a range of adaptations evolved by plants which improve their survival and reproduction by reducing the impact of herbivores. Plants can sense being touched, and they can use several strategies to defend against damage caused by herbivores. Many plants produce secondary metabolites, known as allelochemicals, that influence the behavior, growth, or survival of herbivores. These chemical defenses can act as repellents or toxins to herbivores, or reduce plant digestibility.

Other defensive strategies used by plants include escaping or avoiding herbivores in any time and/or any place, for example by growing in a location where plants are not easily found or accessed by herbivores, or by changing seasonal growth patterns. Another approach diverts herbivores toward eating non-essential parts, or enhances the ability of a plant to recover from the damage caused by herbivory. Some plants encourage the presence of natural enemies of herbivores, which in turn protect the plant. Each type of defense can be either *constitutive* (always present in the plant), or *induced* (produced in reaction to damage or stress caused by herbivores).

Historically, insects have been the most significant herbivores, and the evolution of land plants is closely associated with the evolution of insects. While most plant defenses are directed against insects, other defenses have evolved that are aimed at vertebrate herbivores, such as birds and mammals. The study of plant defenses against herbivory is important, not only from an evolutionary view point, but also in the direct impact that these defenses have on agriculture, including human and livestock food sources; as beneficial 'biological control agents' in biological pest control programs; as well as in the search for plants of medical importance.

Evolution of Defensive Traits

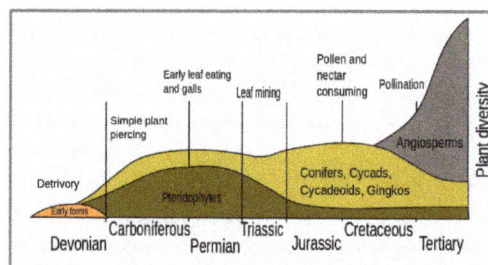

Timeline of plant evolution and the beginnings of different modes of insect herbivory

The earliest land plants evolved from aquatic plants around 450 million years ago (Ma) in the Ordovician period. Many plants have adapted to iodine-deficient terrestrial environment by removing iodine from their metabolism, in fact iodine is essential only for animal cells. An important antiparasitic action is caused by the block of the transport of iodide of animal cells inhibiting sodium-iodide symporter (NIS). Many plant pesticides are glycosides (as the cardiac digitoxin) and cyanogenic glycosides which liberate cyanide, which, blocking cytochrome c oxidase and NIS, is poisonous only for a large part of parasites and herbivores and not for the plant cells in which it seems useful in seed dormancy phase. Iodide is not pesticide, but is oxidized, by vegetable peroxidase, to iodine, which is a strong oxidant, it is able to kill bacteria, fungi and protozoa.

The Cretaceous period saw the appearance of more plant defense mechanisms. The diversification of flowering plants (angiosperms) at that time is associated with the sudden burst of speciation in insects. This diversification of insects represented a major selective force in plant evolution, and led to selection of plants that had defensive adaptations. Early insect herbivores were mandibulate and bit or chewed vegetation; but the evolution of vascular plants lead to the co-evolution of other forms of herbivory, such as sap-sucking, leaf mining, gall forming and nectar-feeding.

The relative abundance of different species of plants in ecological communities including forests and grasslands may be determined in part by the level of defensive compounds in the different species. Since the cost of replacement of damaged leaves is higher in conditions where resources are scarce, it may also be that plants growing in areas where water and nutrients are scarce may invest more resources into anti-herbivore defenses.

Records of Herbivores

Viburnum lesquereuxii leaf with insect damage; Dakota Sandstone (Cretaceous) of Ellsworth County, Kansas. Scale bar is 10 mm.

Our understanding of herbivory in geological time comes from three sources: fossilized plants, which may preserve evidence of defense (such as spines), or herbivory-related damage; the observation of plant debris in fossilised animal faeces; and the construction of herbivore mouthparts.

Long thought to be a Mesozoic phenomenon, evidence for herbivory is found almost as soon as fossils which could show it. Within under 20 million years of the first fossils of sporangia and stems towards the close of the Silurian, around 420 million years ago, there is evidence that they were being consumed. Animals fed on the spores of early Devonian plants, and the Rhynie chert also provides evidence that organisms fed on plants using a "pierce and suck" technique. Many plants of this time are preserved with spine-like enations, which may have performed a defensive role before being co-opted to develop into leaves.

During the ensuing 75 million years, plants evolved a range of more complex organs - from roots to seeds. There was a gap of 50 to 100 million years between each organ evolving, and it being fed upon. Hole feeding and skeletonization are recorded in the early Permian, with surface fluid feeding evolving by the end of that period.

A Plain Tiger *Danaus chrysippus* caterpillar making a moat to block defensive chemicals of *Calotropis* before feeding

Co-evolution

Herbivores are dependent on plants for food, and have evolved mechanisms to obtain this food despite the evolution of a diverse arsenal of plant defenses. Herbivore adaptations to plant defense have been likened to *offensive traits* and consist of adaptations that allow increased feeding and use of a host plant. Relationships between herbivores and their host plants often results in reciprocal evolutionary change, called co-evolution. When an herbivore eats a plant it selects for plants that can mount a defensive response. In cases where this relationship demonstrates *specificity* (the evolution of each trait is due to the other), and *reciprocity* (both traits must evolve), the species are thought to have co-evolved. The "escape and radiation" mechanism for co-evolution presents the idea that adaptations in herbivores and their host plants have been the driving force behind speciation, and have played a role in the radiation of insect species during the age of angiosperms. Some herbivores have evolved ways to hijack plant defenses to their own benefit, by sequestering these chemicals and using them to protect themselves from predators. Plant defenses against herbivores are generally not complete so plants also tend to evolve some tolerance to herbivory.

Types

Plant defenses can be classified generally as constitutive or induced. Constitutive defenses are always present in the plant, while induced defenses are produced or mobilized to the site where a plant is injured. There is wide variation in the composition and concentration of constitutive defenses and these range from mechanical defenses to digestibility reducers and toxins. Many external mechanical defenses and large quantitative defenses are constitutive, as they require large amounts of resources to produce and are difficult to mobilize. A variety of molecular and biochemical approaches are used to determine the mechanism of constitutive and induced plant defenses responses against herbivory.

Induced defenses include secondary metabolic products, as well as morphological and physiological changes. An advantage of inducible, as opposed to constitutive defenses, is that they are only produced when needed, and are therefore potentially less costly, especially when herbivory is variable.

Chemical Defenses

The evolution of chemical defences in plants is linked to the emergence of chemical substances that are not involved in the essential photosynthetic and metabolic activities. These substances, secondary metabolites, are organic compounds that are not directly involved in the normal growth, development or reproduction of organisms, and often produced as by-products during the synthesis of primary metabolic products. Although these secondary metabolites have been thought to play a major role in defenses against herbivores, a meta-analysis of recent relevant studies has suggested that they have either a more minimal (when compared to other non-secondary metabolites, such as primary chemistry and physiology) or more complex involvement in defense.

Persimmon, genus *Diospyros*, has a high tannin content which gives immature fruit, seen above, an astringent and bitter flavor.

Secondary metabolites are often characterized as either *qualitative* or *quantitative*. Qualitative metabolites are defined as toxins that interfere with an herbivore's metabolism, often by blocking specific biochemical reactions. Qualitative chemicals are present in plants in relatively low concentrations (often less than 2% dry weight), and are not dosage dependent. They are usually small, water-soluble molecules, and therefore can be rapidly synthesized, transported and stored with relatively little energy cost to the plant. Qualitative allelochemicals are usually effective against non-adapted specialists and generalist herbivores.

Quantitative chemicals are those that are present in high concentration in plants (5 – 40% dry weight) and are equally effective against all specialists and generalist herbivores. Most quantitative metabolites are digestibility reducers that make plant cell walls indigestible to animals. The effects of quantitative metabolites are dosage dependent and the higher these chemicals' proportion in the herbivore's diet, the less nutrition the herbivore can gain from ingesting plant tissues. Because they are typically large molecules, these defenses are energetically expensive to produce and maintain, and often take longer to synthesize and transport.

The geranium, for example, produces a unique chemical compound in its petals to defend itself from Japanese beetles. Within 30 minutes of ingestion the chemical paralyzes the herbivore. While the chemical usually wears off within a few hours, during this time the beetle is often consumed by its own predators.

See Toxalbumin

Types of Chemical Defences

Plants have evolved many secondary metabolites involved in plant defense, which are collectively known as antiherbivory compounds and can be classified into three subgroups: nitrogen compounds (including *alkaloids, cyanogenic glycosides, glucosinolates* and *benzoxazinoids*), terpenoids, and phenolics.

Alkaloids are derived from various amino acids. Over 3000 known alkaloids exist, examples include nicotine, caffeine, morphine, cocaine, colchicine, ergolines, strychnine, and quinine. Alkaloids have pharmacological effects on humans and other animals. Some alkaloids can inhibit or activate enzymes, or alter carbohydrate and fat storage by inhibiting the formation phosphodiester bonds involved in their breakdown. Certain alkaloids bind to nucleic acids and can inhibit synthesis of proteins and affect DNA repair mechanisms. Alkaloids can also affect cell membrane and cytoskeletal structure causing the cells to weaken, collapse, or leak, and can affect nerve transmission. Although alkaloids act on a diversity of metabolic systems in humans and other animals, they almost uniformly invoke an aversively bitter taste.

Cyanogenic glycosides are stored in inactive forms in plant vacuoles. They become toxic when herbivores eat the plant and break cell membranes allowing the glycosides to come into contact with enzymes in the cytoplasm releasing hydrogen cyanide which blocks cellular respiration. Glucosinolates are activated in much the same way as cyanogenic glucosides, and the products can cause gastroenteritis, salivation, diarrhea, and irritation of the mouth. Benzoxazinoids, secondary defence metabolites, which are characteristic for grasses (Poaceae), are also stored as inactive glucosides in the plant vacuole. Upon tissue disruption they get into contact with β-glucosidases from the chloroplasts, which enzymatically release the toxic aglucones. Whereas some benzoxazinoids are constitutively present, others are only synthesised following herbivore infestation, and thus, considered inducible plant defenses against herbivory.

The terpenoids, sometimes referred to as isoprenoids, are organic chemicals similar to terpenes, derived from five-carbon isoprene units. There are over 10,000 known types of terpenoids. Most are multicyclic structures which differ from one another in both functional groups, and in basic carbon skeletons. Monoterpenoids, continuing 2 isoprene units, are volatile essential oils such as citronella, limonene, menthol, camphor, and pinene. Diterpenoids, 4 isoprene units, are widely distributed in latex and resins, and can be quite toxic. Diterpenes are responsible for making *Rhododendron* leaves poisonous. Plant steroids and sterols are also produced from terpenoid precursors, including vitamin D, glycosides (such as digitalis) and saponins (which lyse red blood cells of herbivores).

Phenolics, sometimes called *phenols*, consist of an aromatic 6-carbon ring bonded to a hydroxy group. Some phenols have antiseptic properties, while others disrupt endocrine activity. Phenolics range from simple tannins to the more complex flavonoids that give plants much of their red, blue, yellow, and white pigments. Complex phenolics called polyphenols are capable of producing many different types of effects on humans, including antioxidant properties. Some examples of phenolics used for defense in plants are: lignin, silymarin and cannabinoids. Condensed tannins, polymers composed of 2 to 50 (or more) flavonoid molecules, inhibit herbivore digestion by binding to consumed plant proteins and making them more difficult for animals to digest, and by interfering with protein absorption and digestive enzymes. Silica and lignins, which are completely indigestible to animals, grind down insect mandibles (appendages necessary for feeding).

In addition to the three larger groups of substances mentioned above, fatty acid derivates, amino acids and even peptides are also used as defense. The cholinergic toxine, cicutoxin of water hemlock, is a polyyne derived from the fatty acid metabolism. β-N-Oxalyl-L-α,β-diaminopropionic acid as simple amino acid is used by the sweet pea which leads also to intoxication in humans. The synthesis of fluoroacetate in several plants is an example of the use of small molecules to disrupt the metabolism of herbivores, in this case the citric acid cycle.

In tropical *Sargassum* and *Turbinaria* species that are often preferentially consumed by herbivorous fishes and echinoids, there is a relatively low level of phenolics and tannins.

Mechanical Defenses

Plants have many external structural defenses that discourage herbivory. Depending on the herbivore's physical characteristics (i.e. size and defensive armor), plant structural defenses on stems and leaves can deter, injure, or kill the grazer. Some defensive compounds are produced internally but are released onto the plant's surface; for example, resins, lignins, silica, and wax cover the epidermis of terrestrial plants and alter the texture of the plant tissue. The leaves of holly plants, for instance, are very smooth and slippery making feeding difficult. Some plants produce gummosis or sap that traps insects.

The thorns on the stem of this raspberry plant, serve as a mechanical defense against herbivory.

A plant's leaves and stem may be covered with sharp prickles, spines, thorns, or trichomes- hairs on the leaf often with barbs, sometimes containing irritants or poisons. Plant structural features like spines and thorns reduce feeding by large ungulate herbivores (e.g. kudu, impala, and goats) by restricting the herbivores' feeding rate, or by wearing down the molars. Raphides are sharp needles of calcium oxalate or calcium carbonate in plant tissues, making ingestion painful, damaging a herbivore's mouth and gullet and causing more efficient delivery of the plant's toxins. The structure of a plant, its branching and leaf arrangement may also be evolved to reduce herbivore impact. The shrubs of New Zealand have evolved special wide branching adaptations believed to be a response to browsing birds such as the moas. Similarly, African Acacias have long spines low in the canopy, but very short spines high in the canopy, which is comparatively safe from herbivores such as giraffes.

Coconut palms protect their fruit by surrounding it with multiple layers of armor.

Trees such as coconut and other palms, may protect their fruit by multiple layers of armor, needing efficient tools to break through to the seed contents, and special skills to climb the tall and relatively smooth trunk.

Thigmonasty

Thigmonastic movements, those that occur in response to touch, are used as a defense in some plants. The leaves of the sensitive plant, *Mimosa pudica*, close up rapidly in response to direct touch, vibration, or even electrical and thermal stimuli. The proximate cause of this mechanical response is an abrupt change in the turgor pressure in the pulvini at the base of leaves resulting from osmotic phenomena. This is then spread via both electrical and chemical means through the plant; only a single leaflet need be disturbed.

This response lowers the surface area available to herbivores, which are presented with the underside of each leaflet, and results in a wilted appearance. It may also physically dislodge small herbivores, such as insects.

Mimicry and Camouflage

Some plants mimic the presence of insect eggs on their leaves, dissuading insect species from laying their eggs there. Because female butterflies are less likely to lay their eggs on plants that already have butterfly eggs, some species of neotropical vines of the genus *Passiflora* (Passion flowers) contain physical structures resembling the yellow eggs of *Heliconius* butterflies on their leaves, which discourage oviposition by butterflies.

Indirect Defenses

The large thorn-like stipules of *Acacia collinsii* are hollow and offer shelter for ants, which in return protect the plant against herbivores.

Another category of plant defenses are those features that indirectly protect the plant by enhancing the probability of attracting the natural enemies of herbivores. Such an arrangement is known as mutualism, in this case of the "enemy of my enemy" variety. One such feature are semiochemicals, given off by plants. Semiochemicals are a group of volatile organic compounds involved in interactions between organisms. One group of semiochemicals are allelochemicals; consisting of allomones, which play a defensive role in interspecies communication, and kairomones, which are used by members of higher trophic levels to locate food sources. When a plant is attacked it releases allelochemics containing an abnormal ratio of these herbivore-induced plant volatiles (HIPVs). Predators sense these volatiles as food cues, attracting them to the damaged plant, and to feeding herbivores. The subsequent reduction in the number of herbivores confers a fitness benefit to the plant and demonstrates the indirect defensive capabilities of semiochemicals. Induced volatiles also have drawbacks, however; some studies have suggested that these volatiles also attract herbivores.

Plants also provide housing and food items for natural enemies of herbivores, known as "biotic" defense mechanisms, as a means to maintain their presence. For example, trees from the genus *Macaranga* have adapted their thin stem walls to create ideal housing for an ant species (genus *Crematogaster*), which, in turn, protects the plant from herbivores. In addition to providing housing, the plant also provides the ant with its exclusive food source; from the food bodies produced by the plant. Similarly, some *Acacia* tree species have developed thorns that are swollen at the base, forming a hollowing structure that acts as housing. These *Acacia* trees also produce nectar in extrafloral nectaries on their leaves as food for the ants.

Plant use of endophytic fungi in defense is a very common phenomenon. Most plants have endophytes, microbial organisms that live within them. While some cause disease, others protect plants from herbivores and pathogenic microbes. Endophytes can help the plant by producing toxins harmful to other organisms that would attack the plant, such as alkaloid producing fungi which are common in grasses such as tall fescue (*Festuca arundinacea*).

Leaf Shedding and Color

There have been suggestions that leaf shedding may be a response that provides protection against diseases and certain kinds of pests such as leaf miners and gall forming insects. Other responses such as the change of leaf colors prior to fall have also been suggested as adaptations that may help undermine the camouflage of herbivores. Autumn leaf color has also been suggested to act as an honest warning signal of defensive commitment towards insect pests that migrate to the trees in autumn.

Costs and Benefits

Defensive structures and chemicals are costly as they require resources that could otherwise be used by plants to maximize growth and reproduction. Many models have been proposed to explore how and why some plants make this investment in defenses against herbivores.

Optimal Defense Hypothesis

The optimal defense hypothesis attempts to explain how the kinds of defenses a particular plant might use reflect the threats each individual plant faces. This model considers three main factors, namely: risk of attack, value of the plant part, and the cost of defense.

The first factor determining optimal defense is risk: how likely is it that a plant or certain plant parts will be attacked? This is also related to the *plant apparency hypothesis*, which states that a plant will invest heavily in broadly effective defenses when the plant is easily found by herbivores. Examples of apparent plants that produce generalized protections include long-living trees, shrubs, and perennial grasses. Unapparent plants, such as short-lived plants of early successional stages, on the other hand, preferentially invest in small amounts of qualitative toxins that are effective against all but the most specialized herbivores.

The second factor is the value of protection: would the plant be less able to survive and reproduce after removal of part of its structure by a herbivore? Not all plant parts are of equal evolutionary value, thus valuable parts contain more defenses. A plant's stage of development at the time of feeding also affects the resulting change in fitness. Experimentally, the fitness value of a plant structure is determined by removing that part of the plant and observing the effect. In general, reproductive parts are not as easily replaced as vegetative parts, terminal leaves have greater value than basal leaves, and the loss of plant parts mid-season has a greater negative effect on fitness than removal at the beginning or end of the season. Seeds in particular tend to be very well protected. For example, the seeds of many edible fruits and nuts contain cyanogenic glycosides such as amygda-lin. This results from the need to balance the effort needed to make the fruit attractive to animal dispersers while ensuring that the seeds are not destroyed by the animal.

The final consideration is cost: how much will a particular defensive strategy cost a plant in energy and materials? This is particularly important, as energy spent on defense cannot be used for other functions, such as reproduction and growth. The optimal defense hypothesis predicts that plants will allocate more energy towards defense when the benefits of protec-tion outweigh the costs, specifically in situations where there is high herbivore pressure.

Carbon: Nutrient Balance Hypothesis

The carbon:nutrient balance hypothesis, also known as the *environmental constraint hypothesis* or *Carbon Nutrient Balance Model* (CNBM), states that the various types of plant defenses are responses to variations in the levels of nutrients in the environment. This hypothesis predicts the Carbon/Nitrogen ratio in plants determines which second-ary metabolites will be synthesized. For example, plants growing in nitrogen-poor soils will use carbon-based defenses (mostly digestibility reducers), while those growing in low-carbon environments (such as shady conditions) are more likely to produce nitro-gen-based toxins. The hypothesis further predicts that plants can change their defenses in response to changes in nutrients. For example, if plants are grown in low-nitrogen conditions, then these plants will implement a defensive strategy composed of consti-tutive carbon-based defenses. If nutrient levels subsequently increase, by for example the addition of fertilizers, these carbon-based defenses will decrease.

Growth Rate Hypothesis

The growth rate hypothesis, also known as the *resource availability hypothesis*, states that defense strategies are determined by the inherent growth rate of the plant, which is in turn determined by the resources available to the plant. A major assumption is that available resources are the limiting factor in determining the maximum growth rate of a plant species. This model predicts that the level of defense investment will increase as the potential of growth decreases. Additionally, plants in resource-poor areas, with inherently slow-growth rates, tend to have long-lived leaves and twigs, and the loss of plant appendages may result in a loss of scarce and valuable nutrients.

A recent test of this model involved a reciprocal transplants of seedlings of 20 species of trees between clay soils (nutrient rich) and white sand (nutrient poor) to determine whether trade-offs between growth rate and defenses restrict species to one habitat. When planted in white sand and protected from herbivores, seedlings originating from clay outgrew those originating from the nutrient-poor sand, but in the presence of her-bivores the seedlings originating from white sand performed better, likely due to their higher levels of constitutive carbon-based defenses. These finding suggest that defen-sive strategies limit the habitats of some plants.

Growth-differentiation Balance Hypothesis

The growth-differentiation balance hypothesis states that plant defenses are a result of a tradeoff between "growth-related processes" and "differentiation-related processes" in different environments. Differentiation-related processes are defined as "processes that enhance the structure or function of existing cells (i.e. maturation and specialization)." A plant will produce chemical defenses only when energy is available from photosynthesis, and plants with the highest concentrations of secondary metabolites are the ones with an intermediate level of available resources. The GDBH also accounts for tradeoffs between growth and defense over a resource availability gradient. In situations where resources (e.g. water and nutrients) limit photosynthesis, carbon supply is predicted to limit both growth and defense. As resource availability increases, the requirements needed to support photosynthesis are met, allowing for accumulation of carbohydrate in tissues. As resources are not sufficient to meet the large demands of growth, these carbon compounds can instead be partitioned into the synthesis of carbon based secondary metabolites (phenolics, tannins, etc.). In environments where the resource demands for growth are met, carbon is allocated to rapidly dividing meristems (high sink strength) at the expense of secondary metabolism. Thus rapidly growing plants are predicted to contain lower levels of secondary metabolites and vice versa. In addition, the tradeoff predicted by the GDBH may change over time, as evidenced by a recent study on Salix spp. Much support for this hypothesis is present in the literature, and some scientists consider the GDBH the most mature of the plant defense hypotheses.

Importance to Humans

Agriculture

The variation of plant susceptibility to pests was probably known even in the early stages of agriculture in humans. In historic times, the observation of such variations in susceptibility have provided solutions for major socio-economic problems. The grape phylloxera was introduced from North America to France in 1860 and in 25 years it destroyed nearly a third (100,000 km²) of the French grape yards. Charles Valentine Riley noted that the American species *Vitis labrusca* was resistant to *Phylloxera*. Riley, with J. E. Planchon, helped save the French wine industry by suggesting the grafting of the susceptible but high quality grapes onto *Vitis labrusca* root stocks. The formal study of plant resistance to herbivory was first covered extensively in 1951 by Reginald (R.H.) Painter, who is widely regarded as the founder of this area of research, in his book *Plant Resistance to Insects*. While this work pioneered further research in the US, the work of Chesnokov was the basis of further research in the USSR.

Fresh growth of grass is sometimes high in prussic acid content and can cause poisoning of grazing livestock. The production of cyanogenic chemicals in grasses is primarily a defense against herbivores.

The human innovation of cooking may have been particularly helpful in overcoming many of the defensive chemicals of plants. Many enzyme inhibitors in cereal grains and pulses, such as trypsin inhibitors prevalent in pulse crops, are denatured by cooking, making them digestible.

It has been known since the late 17th century that plants contain noxious chemicals which are avoided by insects. These chemicals have been used by man as early insecticides; in 1690 nicotine was extracted from tobacco and used as a contact insecticide. In 1773, insect infested plants were treated with nicotine fumigation by heating tobacco and blowing the smoke over the plants. The flowers of *Chrysanthemum* species contain pyrethrin which is a potent insecticide. In later years, the applications of plant resistance became an important area of research in agriculture and plant breeding, particularly because they can serve as a safe and low-cost alternative to the use of pesticides. The important role of secondary plant substances in plant defense was described in the late 1950s by Vincent Dethier and G.S. Fraenkel. The use of botanical pesticides is widespread and notable examples include Azadirachtin from the neem (*Azadirachta indica*), d-Limonene from Citrus species, Rotenone from *Derris*, Capsaicin from chili pepper and Pyrethrum.

Natural materials found in the environment also induce plant resistance as well. Chitosan derived from chitin induce a plant's natural defense response against pathogens, diseases and insects including cyst nematodes, both are approved as biopesticides by the EPA to reduce the dependence on toxic pesticides.

The selective breeding of crop plants often involves selection against the plant's intrinsic resistance strategies. This makes crop plant varieties particularly susceptible to pests unlike their wild relatives. In breeding for host-plant resistance, it is often the wild relatives that provide the source of resistance genes. These genes are incorporated using conventional approaches to plant breeding, but have also been augmented by recombinant techniques, which allow introduction of genes from completely unrelated organisms. The most famous transgenic approach is the introduction of genes from the bacterial species, *Bacillus thuringiensis*, into plants. The bacterium produces proteins that, when ingested, kill lepidopteran caterpillars. The gene encoding for these highly toxic proteins, when introduced into the host plant genome, confers resistance against caterpillars, when the same toxic proteins are produced within the plant. This approach is controversial, however, due to the possibility of ecological and toxicological side effects.

Pharmaceutical

Many currently available pharmaceuticals are derived from the secondary metabolites plants use to protect themselves from herbivores, including opium, aspirin, cocaine, and atropine. These chemicals have evolved to affect the biochemistry of insects in very specific ways. However, many of these biochemical pathways are conserved in vertebrates, including humans, and the chemicals act on human biochemistry in ways similar to that of insects. It has therefore been suggested that the study of plant-insect interactions may help in bioprospecting.

Illustration from the 15th-century manuscript *Tacuinum Sanitatis* detailing the beneficial and harmful properties of Mandrakes

There is evidence that humans began using plant alkaloids in medical preparations as early as 3000 B.C. Although the active components of most medicinal plants have been isolated only recently (beginning in the early 19th century) these substances have been used as drugs throughout the human history in potions, medicines, teas and as poisons. For example, to combat herbivory by the larvae of some Lepidoptera species, Cinchona trees produce a variety of alkaloids, the most familiar of which is quinine. Quinine is extremely bitter, making the bark of the tree quite unpalatable, it is also an anti-fever agent, known as Jesuit's bark, and is especially useful in treating malaria.

Throughout history mandrakes (*Mandragora officinarum*) have been highly sought after for their reputed aphrodisiac properties. However, the roots of the mandrake plant also contain large quantities of the alkaloid scopolamine, which, at high doses, acts as a central nervous system depressant, and makes the plant highly toxic to herbivores. Scopolamine was later found to be medicinally used for pain management prior to and during labor; in smaller doses it is used to prevent motion sickness. One of the most well-known medicinally valuable terpenes is an anticancer drug, taxol, isolated from the bark of the Pacific yew, *Taxus brevifolia*, in the early 1960s.

Biological Pest Control

Repellent companion planting, defensive live fencing hedges, and "obstructive-repellent" interplanting, with host-plant resistance species as beneficial 'biological control agents' is a technique in biological pest control programs for: organic gardening, wildlife gardening, sustainable gardening, and sustainable landscaping; in organic farming and sustainable agriculture; and in restoration ecology methods for habitat restoration projects.

References

- Futuyma, Douglas J.; Montgomery Slatkin (1983). Coevolution. Sunderland, Massachusetts: Sinauer Associates. ISBN 0-87893-228-3.

- Whittaker, Robert H. (1970). "The biochemical ecology of higher plants". In Ernest Sondheimer; John B. Simeone. Chemical ecology. Boston: Academic Press. pp. 43–70. ISBN 0-12-654750-5.

- Roberts, Margaret F.; Michael Wink (1998). Alkaloids: biochemistry, ecology, and medicinal applications. New York: Plenum Press. ISBN 0-306-45465-3.

- Raven, Peter H.; Ray F. Evert; Susan E. Eichhorn (2005). Biology of Plants. New York: W. H. Freeman and Company. ISBN 0-7167-1007-2.

- Krischik, V. A.; R. F. Denno (1983). "Individual, population, and geographic patterns in plant defense.". In Robert F. Denno; Mark S. McClure. Variable plants and herbivores in natural and managed systems. Boston: Academic Press. pp. 463–512. ISBN 0-12-209160-4.

- Michael Smith, C. (2005). Plant Resistance to Arthropods: Molecular and Conventional Approaches. Berlin: Springer. ISBN 1-4020-3701-5.

- Chapin, F. Stuart, III (1980). "The Mineral Nutrition of Wild Plants". Annual Review of Ecology and Systematics. 11: 233–260. doi:10.1146/annurev.es.11.110180.001313. JSTOR 2096908. Retrieved 2014-01-15

- Venturi, Sebastiano (2011). "Evolutionary Significance of Iodine". Current Chemical Biology-. 5 (3): 155–162. doi:10.2174/187231311796765012. ISSN 1872-3136.

- Carmona, Diego; Marc J. Lajeunesse; Marc T.J. Johnson (April 2011). "Plant traits that predict resistance to herbivores" (PDF). Functional Ecology. 25 (2): 358–367. doi:10.1111/j.1365-2435.2010.01794.x. Retrieved 26 June 2011.

Permissions

All chapters in this book are published with permission under the Creative Commons Attribution Share Alike License or equivalent. Every chapter published in this book has been scrutinized by our experts. Their significance has been extensively debated. The topics covered herein carry significant information for a comprehensive understanding. They may even be implemented as practical applications or may be referred to as a beginning point for further studies.

We would like to thank the editorial team for lending their expertise to make the book truly unique. They have played a crucial role in the development of this book. Without their invaluable contributions this book wouldn't have been possible. They have made vital efforts to compile up to date information on the varied aspects of this subject to make this book a valuable addition to the collection of many professionals and students.

This book was conceptualized with the vision of imparting up-to-date and integrated information in this field. To ensure the same, a matchless editorial board was set up. Every individual on the board went through rigorous rounds of assessment to prove their worth. After which they invested a large part of their time researching and compiling the most relevant data for our readers.

The editorial board has been involved in producing this book since its inception. They have spent rigorous hours researching and exploring the diverse topics which have resulted in the successful publishing of this book. They have passed on their knowledge of decades through this book. To expedite this challenging task, the publisher supported the team at every step. A small team of assistant editors was also appointed to further simplify the editing procedure and attain best results for the readers.

Apart from the editorial board, the designing team has also invested a significant amount of their time in understanding the subject and creating the most relevant covers. They scrutinized every image to scout for the most suitable representation of the subject and create an appropriate cover for the book.

The publishing team has been an ardent support to the editorial, designing and production team. Their endless efforts to recruit the best for this project, has resulted in the accomplishment of this book. They are a veteran in the field of academics and their pool of knowledge is as vast as their experience in printing. Their expertise and guidance has proved useful at every step. Their uncompromising quality standards have made this book an exceptional effort. Their encouragement from time to time has been an inspiration for everyone.

The publisher and the editorial board hope that this book will prove to be a valuable piece of knowledge for students, practitioners and scholars across the globe.

Index

www.ingramcontent.com/pod-product-compliance
Lightning Source LLC
Chambersburg PA
CBHW061938190326
41458CB00009B/2766